Advanced Information and Knowledge Processing

Series Editors
Professor Lakhmi Jain
Lakhmi.jain@unisa.edu.au

Professor Xindong Wu
xwu@cs.uvm.edu

For other titles published in this series, go to
http://www.springer.com/series/4738

Maria Chli · Philippe de Wilde

Convergence and Knowledge Processing in Multi-Agent Systems

Springer

Maria Chli
Aston University
Birmingham
United Kingdom B4 7ET

Philippe de Wilde
Heriot Watt University
Edinburgh
United Kingdom EH14 4AS

ISBN 978-1-84882-062-3 e-ISBN 978-1-84882-063-0
DOI 10.1007/978-1-84882-063-0
Springer Dordrecht Heidelberg London New York

British Library Cataloguing in Publication Data
A catalogue record for this book is available from the British Library

Library of Congress Control Number: 2009926287

Cover design: SPi Publisher Services

Printed on acid-free paper

Springer is part of Springer Science+Business Media (www.springer.com)

Preface

Agent systems are being used to model complex systems like societies, markets and biological systems. Our broader interest lies in understanding the dynamics as well as in analysing the properties which emerge from the interactions that occur in such systems. More specifically, in this work we investigate issues of agent systems related to convergence and interactivity. We have been using techniques from agent-based modelling to simulate complex systems. The work described in this book makes four main contributions to the fields of convergence and knowledge processing in agent systems.

First, we propose a definition for the stability of multi-agent systems. The system is perceived as a discrete time Markov chain with a potentially unknown transition probability distribution. It is considered to be stable when its state has converged to an equilibrium distribution. The definition proposed is the only one which takes into account the game nature of multi-agent systems, is relevant to systems with a varying number of agents and is supported by the mathematical framework of stochastic systems. Several artificial ecosystems have been implemented and used to verify the proposed definition and carry out an analysis of the stability of multi-agent systems.

Second, we investigate knowledge exchange among agents in a market scenario. The forces that drive it are identified as well as its effects on the overall behaviour, and especially the convergence of the system. Knowledge exchange is known to be beneficial for industry, but in order to explain it, authors have used high-level concepts like network effects, reputation and trust. Even though the model we present does not include any such concepts, information exchange naturally emerges as a successful, profitable behaviour. This behaviour is shown to increase the efficiency of the market.

Third, we show how information provided through interaction of users in a scenario can be used to optimise the queries submitted. The proposed algorithm is based on the observation that documents are often found to contain terms with high information content which can summarise their subject matter. Experiments carried out demonstrate that our approach significantly shortens the web search sessions as well as the number of documents viewed.

Fourth, we describe a pricing strategy for a realistic large-scale distributed system. This system consists of automatic personal assistants (PAs) that can book time slots in each other's diaries, and that have to pay for doing so. Stability of strategies is first studied in an evolutionary context early in this book. We call the strategy stable if it prevents deadlock in the network, when none of the PAs buys or sells resources anymore. A stable strategy is robust if substantial noise on the parameters and the initial conditions maintains stability.

Empirical results validate all four contributions within a number of domains. The generality of the contributions is verified by applying them to simulations of complex market, social and biological systems. Ultimately, this book sheds light upon the complex interrelation between interactivity/exchange of knowledge and convergence in multi-agent systems.

We would like to thank all our collaborators in the EEII (Appendix A) and DBE (Section 4.2) projects for the inspiration and the useful discussions. The second author would like to thank Drs. Nader Azarmi, Ben Azvine, and Lyndon Lee at British Telecom (BT) for a collaboration that spanned several years, and for suggesting many inspiring problems. Their promotion of multi-agent systems and the need for analysis of such systems has been forward-looking.

Birmingham and Edinburgh, UK, *Maria Chli*
March 2009 *Philippe De Wilde*

Contents

List of Figures

List of Tables

Acronyms

ACE Agent-based computational economics

DBE Digital business ecosystem; more details are available in Section 4.2.

DoF Degrees of freedom

EEII Evolution and ecology of interacting infohabitants; more details are available in Appendix A.

GA Genetic algorithm

GPS Global positioning system

MAS Multi-agent system

PA Personal assistant

R&D Research and development

SME Small and medium enterprises

Chapter 1
Introduction

Intelligent agents are software entities that carry out a set of operations on behalf of a user or another program. They operate with a certain degree of autonomy, employing some knowledge of the user's goals. Multi-agent systems are complex systems comprised of multiple intelligent agents that act either independently or in cooperation with one another.

According to Jennings et al. [1998], multi-agent systems share a number of key characteristics: (a) each agent has incomplete information or capabilities for solving the problem and, thus has a limited viewpoint; (b) there is no system global control; (c) data is decentralised; and (d) computation is asynchronous. The aforementioned characteristics make multi-agent systems a suitable method for modelling, simulation and analysis of real-life complex systems. In recent years, many researchers have contributed valuable work in multi-agent systems, which is applicable to many disciplines including economics [Kaihara, 2001], sociology [Rodrigues et al., 2003], organisation and management science [Huin, 2004] and biology [Guo et al., 2005].

In complex systems such as organisms, ecosystems, markets or societies the issue of *convergence* is often important. It is a fairly broad notion, and its exact meaning might vary in each context. Researchers, [Lee et al., 1998; De Wilde et al., 1999a; Bredin et al., 2000; Schillo et al., 2001], have been investigating whether the systems they study converge, and under which circumstances.

Interaction, and in particular cooperation through knowledge exchange, among agents in a multi-agent system is another central and widely studied problem. Interaction can take place in a multitude of ways [Doran et al., 1997], having a significant impact on the overall behaviour of the system.

The objective of this book is to understand the dynamics and analyse the properties which emerge from the interactions that occur in multi-agent systems. In particular, we study the effect agent interactions have on the convergence of a system using agent-based modelling to simulate complex systems. Multi-agent systems offer strong models for representing real-world environments with an appropriate degree of complexity and dynamism [Luck and McBurney, 2005]. Simulation of economies, societies and biological environments are typical application areas. The work described in this book can be summarised in four main contributions:

M. Chli and P. De Wilde, *Convergence and Knowledge Processing in Multi-Agent Systems*,
Advanced Information and Knowledge Processing,

1. Analysis of stability of multi-agent systems and proposal of a formal definition for this notion.
2. Study of the necessary conditions which give rise to exchange among agents in an economy as well as of the effect exchanging information has on the efficiency of a market .
3. Investigation of cooperation among users in a scenario. Proposal of an algorithm which enables searchers to interact in order to enhance their queries.
4. Development of a computational model of a realistic large-scale distributed system which carries out resource management tasks.

1.1 Background to the Research

This book investigates the inter-relation between agent interaction, knowledge exchange and system convergence in multi-agent systems. While not many researchers have studied the two issues in conjunction, there has been previous work examining them independently. A more complete review of the related research can be found in Chapter 2. There have been several approaches to examining as well as defining the notion of convergence in multi-agent systems. In De Wilde et al. [1999a] and Lee et al. [1998] the stability of such systems has been investigated within the framework of stochastic processes. The related notion of robustness has been defined in Schillo et al. [2001] in terms of the effect several perturbations incur on the performance of the system. Another common treatment of system convergence has been the game theoretic one [Bredin et al., 2000] where the system convergence is associated with the Nash equilibrium.

Agent interaction can occur in various ways. They range from a "streamline" type of cooperation where an agent completes its task before handing over to the next agent as in Moukas and Maes [1998], to emergent cooperation in which the agents appear to be working together, whereas in fact they act independently as is the case in ant colonies [Dorigo and Gambardella, 1997], to more complex types of cooperation which involve coordination or negotiation [Conlon, 2003; Vidal, 2002; Li et al., 2006].

The goal of the research presented in this book is to illustrate emergent converging behaviour of a complex system from the interactions of agents which follow simple strategies.

1.2 Approach

Agent-based modelling has been used to simulate various scenarios of complex systems including load transportation, virus spreading and market scenarios. The systems have been modelled from an ecological viewpoint. In other words, there are no central controlling powers and all properties of the system arise from distributed interactions. The agents are lightweight, with bounded rationality and no complex reasoning capabilities. In many cases the simulation results are analysed with the aid of statistical analysis.

1.3 Contributions

The work described in this book makes four contributions.

1. We investigate the fundamental concepts behind the stability of multi-agent systems. We view the system as a discrete time Markov chain with a potentially unknown transition probability distribution. The system will be considered to be stable when its state has converged to an equilibrium distribution. The definition proposed is the only one which takes into account the game nature of multi-agent systems, is relevant to systems with a varying number of agents and is supported by the mathematical framework of stochastic systems. We describe some artificial multi-agent ecosystems that were developed and we present results based on these systems. The results confirm that this approach qualitatively agrees with our intuition.
2. We investigate knowledge exchange among agents, the rationale behind it and its effects on the ecosystem. This investigation is performed in a market setting. Knowledge exchange is known to be beneficial for industry, but in order to explain it, authors have used high-level concepts like network effects, reputation and trust. We attempt to formalise a plausible and elegant explanation of how and why organisations adopt information exchange and why it benefits the market as a whole when this happens. This explanation is based on a multi-agent model that simulates a market of software providers. Even though the model does not include any high-level concepts, information exchange naturally emerges during simulations as a successful profitable behaviour. The conclusions reached by this agent-based analysis are twofold: (1) A straightforward set of assumptions is enough to give rise to exchange in a market. (2) Knowledge exchange is shown to increase the efficiency of the market.
3. We investigate how information provided through interaction with the system can be used to optimise its performance. The results of this work are illustrated in the context of the Internet retrieval process. The expansion of the Internet has made the task of searching a crucial one. Internet users, however, have to make a great effort in order to formulate a search query that returns the required results. Many methods have been devised to assist in this task by helping the users modify their query to give better results. In this work we propose an interactive method for query expansion. It is based on two observations: (1) documents are often found to contain terms with high information content which can summarise their subject matter and (2) some users are more proficient in formulating accurate queries than others. We present experimental results which demonstrate that our approach significantly shortens the time required in order to accomplish a certain task by performing web searches.

4. To the examples developed in Chapters 3 and 4 — trading, load transportation, virus spreading and software services — we add a fifth one, the trading of time slots associated with staff and material resources. This combines resource management with scheduling . We follow an agent-based approach, and study convergence. We develop a computational model of the resource management tasks of a general personal assistant. The scheduling of time slots is crucial in this. By assigning values to resources , we propose a micro-economic trading strategy for the PAs. This is a free market strategy with inflation and discount on bankruptcy. We simulate the network of PAs using this strategy, with a random stream of incoming requests for time slots. We find that the strategy is stable and robust. We have developed an automatic multi-agent-based system that can take on most of the resource and scheduling tasks of a personal assistant.

This work demonstrates the fact that interaction and convergence in multi-agent systems are significantly interrelated. Simulations applied to a number of domains in market and social scenarios confirm the validity of all four contributions.

1.4 Reader's Guide to the Book

The remainder of this book is structured as follows:

Chapter 2 explores research issues related to the work described in this book presenting previous approaches to the problems discussed. Sections covering more specific background information are included in each of the chapters 3, 4 and 6.

Chapter 3 presents the definition of stability we propose and shows how it is applied to a set of simple games as well as more complex multi-agent systems. Part of the work described in this chapter has been published in Chli et al. [2003] and De Wilde et al. [2003].

Chapter 4 investigates information exchange in a market scenario. The minimal set of assumptions required to give rise to knowledge exchange among competitors in a market is studied. Additionally, in this chapter, it is shown that these interactions lead to higher market efficiency levels. This work is included in Chli and De Wilde [2009].

Chapter 5 proposes an algorithm for collaboration among users in a context. Experimental results testify the effectiveness of the approach in shortening the search sessions. Part of the work described in this chapter is published in Chli and De Wilde [2005].

Chapter 6 describes a stable strategy for a realistic large-scale distributed system. This system consists of automatic personal assistants (PAs) that can book timeslots in each other's diaries. Stable strategies for PAs are analysed in the light of the definitions in Chapter 3. A computational model of the resource management tasks of a general PA is developed.

Chapter 7 concludes the book by summarising what has been achieved and pointing directions for further work.

Chapter 2
Research Issues

Abstract In this chapter, previous work as well as several general research issues that are connected to the work described in the chapters to follow are discussed. Additionally, relevant background work is reviewed in more detail in each of the following chapters separately.

2.1 Multi-agent Systems

Multi-agent systems are computational systems in which several semi-autonomous, intelligent agents, such as software programs or robots, interact or work together to perform some set of tasks or satisfy some set of goals [Lesser, 1995]. Their interactions can be either cooperative or selfish. That is, the agents can share a common goal (as in an ant colony), or they can pursue their own interests (as in the free market economy).

In Maes [1994], intelligent agents were defined as software entities, that inhabit some complex dynamic environment. Within that environment, they carry out some set of operations on behalf of a user or another program with some degree of independence or autonomy, and in so doing, employ some knowledge or representation of the user's goals or desires.

Most work in multi-agent systems can be roughly divided into two classes: (a) research concerned with modelling of complex systems and performing simulations (e.g. SWARM simulations [Daniels, 1999]), and (b) research concerned with investigation of real-life multi-agent systems, their design and deployment as well as agent strategies.

In the first class, multi-agent systems simulations have been used in many disciplines. Some examples include modelling immune systems [Guo et al., 2005] in biology as well as modelling social exchanges among individuals in societies in the discipline of sociology [Rodrigues et al., 2003]. In economics, multi-agent systems have been used to model markets and supply chains as in Kaihara [2001] and in organisation and management science in modelling enterprise resources planning

M. Chli and P. De Wilde, *Convergence and Knowledge Processing in Multi-Agent Systems*, Advanced Information and Knowledge Processing,

and project management [Huin, 2004]. Agent-based modelling is reviewed in more detail in Section 2.2 below.

In the second class, research concerned with investigation of real-life multi-agent systems, researchers engage in managing the ways in which the agents interact. For example, they have developed agent communication languages [Horling and Lesser, 2005] and protocols for agent interaction that are suitable for different settings [Shen et al., 2005] (e.g. auctions, negotiation). Researchers have been recently utilising argumentation [McBurney and Parsons, 2004] theory, the philosophy of argument and debate, to design richer languages for agent interaction which are able to support argument and nondeductive reasoning. There has also been significant work proposing several agent architectures [D'Inverno et al., 2004] to facilitate the development of real-life multi-agent systems. In this class, a very active research area is the investigation of issues of trust and reputation [Ashri et al., 2005] among the agents within an electronic marketplace based either on using a history of interactions or on recommendations from other agents. Other very active research areas are the development of strategies for different types of automated negotiation [Nguyen and Jennings, 2005; Li et al., 2006] and auction [Fatima et al., 2005; Likhodedov and Sandholm, 2005].

2.2 Agent-based Modelling

Agent-based modelling according to Tesfatsion [2005] "is a method for studying systems exhibiting the following two properties:

1. the system is composed of interacting agents; and
2. the system exhibits *emergent* properties, that is, properties arising from the interactions of the agents that cannot be deduced simply by aggregating the properties of the agents."

Agent-based modelling has been recently used in economics research work to study markets and their characteristics [Kirman and Vriend, 2001], in computing-economics interdisciplinary work to study information economies of autonomous agents [Kephart et al., 1997], in Social Sciences to study emergent behaviour [Epstein, 2002] and in other disciplines.

The issue of how agents form interaction networks is addressed by [Kirman and Vriend, 2001] in the context of agent-based computation. This work presents a model of the actual wholesale fish market in Marseilles, France. Two features characterising this market are: (a) loyalty relationships between particular buyers and sellers; and (b) persistent price dispersion unexplainable by observable characteristics of the fish. Although the model does not make any assumptions that would result in these two characteristic features to be present in the simulation results, loyalty relationships indeed emerge naturally between particular buyer-seller pairs as the buyers and sellers co-evolve their trading rules over time. Buyers learn to become loyal to particular sellers while, at the same time, sellers learn to offer higher payoffs to their more loyal buyers. Moreover, this evolving trade network supports persistent price dispersion over time.

A scenario of retailing information is presented in Kephart et al. [1997]. Broker agents gather information from producers and distribute it to the consumers. The dynamical behaviour of the brokers, and especially their price-setting mechanisms, are studied in the context of a simple information-filtering economy. The broker agents follow a simple price-setting strategy based solely on the system's current state, without prediction of the future. Results show that the system's dynamical behaviour in such "myopic" cases results in an unending cycle of competitive "wars" in the price/product space.

In Epstein [2002] a spatial agent-based model is used to explore civil revolution. A central authority uses "cops" to arrest (remove) actively rebelling citizens from the society for a specified jail term. In each time step, each agent (cop or citizen) randomly moves to a new unoccupied site within its limited vision. A rebelling citizen's estimated arrest probability is assumed to depend on the ratio of actively rebelling citizens to cops that the citizen perceives in its vicinity. Each citizen in each time step decides whether to actively rebel or not depending on this perceived ratio as well as on the hardship and "illegitimacy" of the citizen. It is shown how the complex dynamics resulting from these simple assumptions can generate empirically interesting macroscopic regularities that are difficult to analyze using more standard modelling approaches.

In models like those presented in the chapters to follow, where the interaction of the agents leads to emergent phenomena, mathematical analysis is typically very limited in its ability to derive the dynamic consequences. For this reason, agent-based modelling and simulation are deemed practical methods for analysis of the arising properties of the systems under investigation.

2.3 An Ecosystem Perspective of Multi-agent Systems

The notion of an ecosystem originates from biology and can be defined as an entity composed of *living* organisms *interacting* with each other and with the environment in which they inhabit. Central controlling powers do not exist in ecosystems, all properties of the system arise from distributed interactions. Although the capability of each organism on its own may be very simple, the collective behaviours emerging from their interactions exceed the capacities of any individual organism [Van Dyke Parunak et al., 1997].

An ecosystem-inspired approach provides to the researcher the ability to design decentralised and dynamic systems that are suitable for modelling real-life situations. Ecosystems have been a source of inspiration to a number of developers of agent systems [Moukas and Maes, 1998; Van Dyke Parunak et al., 1997; Bousquet and Page, 2004].

Ecosystem inspired ideas are employed in the Amalthaea architecture presented in Moukas and Maes [1998] for information filtering. The system is populated with two different categories of agents: filtering agents which model and monitor the interests of a user and discovery agents which model the information sources. The agents evolve, compete, and collaborate in a market-like ecosystem. Agents that are useful to the user or other agents are allowed to reproduce, while low-performing agents are destroyed.

A “synthetic ecosystems” approach, which uses the interaction of many simple agents to solve problems, is proposed in Van Dyke Parunak et al. [1997] for the design of multi-agent systems. The main focus is on defining the specification of such ecosystems as well as the types of interactions among the agents. This approach is contrasted to that of sophisticated processing at the individual agent level.

A review of ecosystem-inspired agent systems which use agent-based models to assist in the management of natural ecosystems is presented in Bousquet and Page [2004]. It is argued that researchers in the field of ecosystem management can use multi-agent systems to go beyond the role of the individual and to study more deeply and more effectively the different forms of organisation (spatial, networks, hierarchies) in a natural ecosystem as well as interactions among different organisational levels.

The design of the multi-agent systems presented in the chapters to follow is based on ecology principles. Agents share the environment resources and interact with each other in a way that loosely parallels the relationship of biological organisms and their natural environment. The agents implemented are lightweight and of bounded rationality in a way similar to that presented in Sandholm and Lesser [1997]. However, the cost of computation is not explicitly taken into account as is the case for Sandholm and Lesser [1997]. The agents are lightweight in the sense that they do not exhibit any complex behaviour on their own and they do not have complete information about the state of the system. Agents may exchange information in order to realise their objectives. The dynamics of these ecosystem-inspired multi-agent systems and their emerging properties are investigated.

2.4 Convergence Issues

In this section we review several approaches to convergence of multi-agent systems. As the notion of convergence is fairly broad, several diverse treatments of the issue exist. The papers discussed below are representative samples of the different approaches. In De Wilde et al. [1999a] a multi-agent system scenario is presented in which the agents execute tasks that produce resources. The agent system can be abstracted as a stochastic Markov process. Stability, is regarded as a property of an equilibrium as in Lee et al. [1998]. Although a concrete definition is not given it is intuitively argued that a multi-agent system is in equilibrium when the statistical properties of its performance indicators remain stationary for a given variation in the system's external load. For example, in De Wilde et al. [1999a] a system is deemed stable when the proportion of agents that execute tasks is stationary.

A notion related to stability is that of robustness of multi-agent systems. In Schillo et al. [2001] robustness is quantitatively defined by the expected change of the performance measure in four perturbation scenarios, namely (a) increase of population size, (b) change of task profile over time, (c) malicious agent intrusion, and (d) drop-outs of agents.

Another approach to convergence in the multi-agent system context is the game theoretic one. Many researchers associate system convergence with Nash equilibrium. For example, in Bredin et al. [2000] the problem of resource allocation in a network of mobile agents competing for computational priority is considered. The problem is formulated as a multi-agent game with the players being agents purchasing service from a common server. The system is shown to reach a Nash equilibrium under certain conditions.

2.5 Interaction and Knowledge Exchange

Agent interaction is often presented as one of the key concepts which differentiates multi-agent systems from other related disciplines such as distributed computing, object-oriented systems, and expert systems. Franklin [Doran et al., 1997] (see Figure 2.1) discriminates between different types of agent interaction/cooperation in the context of multi-agent systems.

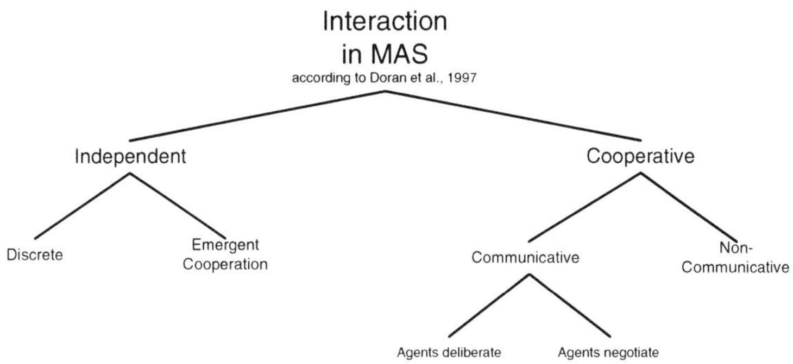

Fig. 2.1: Interaction in MAS according to Franklin [Doran et al., 1997].

According to Franklin's classification a multi-agent system can be either independent or cooperative. A system is classified as independent when each of its agents pursues its own agenda independently of the other agents in the system. Independent systems are subdivided further to discrete systems and systems that exhibit emergent cooperation. In a discrete system the agendas of the agents bear no relation to one another. Discrete systems involve no cooperation between the agents. An example of a discrete multi-agent system is the Amalthaea [Moukas and Maes, 1998] where two general species of agents exist: information filtering agents and information discovery agents. The former are responsible for the personalisation of the system and for keeping track of the interests of the user. The latter are responsible for information resources handling and finding the actual information that the user is interested in. The agents have no point of contact with one another.

In systems that exhibit emergent cooperation agents cooperate with no intention of doing so. Ant colonies, like the ones described in Dorigo and Gambardella [1997], are instances of such systems. Ants search for food sources in their habitat, independently depositing pheromone along their paths. The better the path, the stronger the pheromone deposition along its trail. In this manner, pheromones accrue more along optimal routes. Ant movements in turn are governed by pheromone concentrations, so that the search is confined to better regions of the search space. Cooperation is an emergent behaviour of the system in that, from an observer's viewpoint, the agents appear to be working together, but from the agent's viewpoint they are not. They are simply carrying out their own individual behaviour.

A review of systems that use biologically inspired design methodologies and exhibit emergent cooperative behaviour can be found in Nitschke [2005]. The research results reviewed in that work were from three problem domains, those of swarm-based systems, pursuit and evasion systems and RoboCup Soccer. Swarm-based systems include simulation models of social organisations of ant colonies, like the model described in Dorigo and Gambardella [1997], as well as swarm-bots. Swarm-bots are systems concerned with applying biologically inspired design principles in the simulation of groups of mobile robots that exploit concepts such as self-assembly and self-organisation in order to accomplish collective goals, e.g. [Mondada et al., 2002]. Pursuit-evasion systems deal with scenarios in which a team of pursuers try to capture a group of evaders within a certain environment, e.g. [Vidal et al., 2002]. The goal of research in pursuit-evasion systems is often to illustrate emergent cooperative behaviour from the interactions of pursuers following simple pursuit strategies. Finally, RoboCup Soccer is a field of research dedicated to the design of multi-robot systems for playing a robot form of soccer. Note that not all systems reviewed in Nitschke [2005] strictly comply with Franklin's [Doran et al., 1997] definition of systems that exhibit emergent cooperation. Nevertheless, an important point which is made in the review is that in most cases, in all three problem domains, the mechanisms responsible for the observed cooperative behaviour were not clearly identified by the researchers. In other words, the application of biologically inspired design principles to current artificial social system simulations lacks proven methodologies that allow for effective analysis and evaluation of the mechanisms that motivate the emergence of cooperation.

The complement of independent systems are those in which the agendas of the agents include cooperating with other agents in some way, creating cooperative systems. Such cooperation can either be communicative in that the agents communicate through the intentional sending and receiving of signals with each other, or it can be non-communicative. In the latter case, agents coordinate their cooperative activity by each observing and reacting to the behaviour of others, as in a repeated prisoner's dilemma situation [Conlon, 2003]. Communicative systems can take two forms; agents can deliberate or they can negotiate. In deliberative systems agents jointly plan their actions so as to cooperate with each other. Such a system is described in Vidal [2002] where agents coordinate in order to find optimal solutions in task-oriented domains. Negotiating systems, such as [Li et al., 2006], are like deliberative systems, except that they have an added dose of competition.

Chapter 3
Stability of Multi-agent Systems

Abstract This chapter attempts to investigate the fundamental concepts behind the stability of multi-agent systems. We view the system as a discrete-time Markov chain with potentially unknown transition probability distribution. The system will be considered to be stable when it has converged to an equilibrium distribution. Faced with the nontrivial task of establishing the convergence to such a distribution, we propose a hypothesis testing approach according to which we test whether the convergence of a particular system metric has occurred. We describe some artificial multi-agent ecosystems that were developed for analysis and verification purposes. We present results based on these multi-agent systems that confirm that the proposed approach qualitatively agrees with our intuition. [1]

3.1 Introduction

In this chapter we focus on the concept of stability, with regards to multi-agent systems. Having examined the concept of stability in real-world systems, we propose a definition that is supported by the mathematical framework of stochastic systems. Consecutively, we present the most representative of the experiments we carried out on different multi-agent systems, and apply our definition to test whether they are stable or not.

We are interested in understanding the dynamics of agent interactions [De Wilde et al., 1999a, 2003; Chli et al., 2003]. This effort fits within the framework of the Evolution and Ecology of Interacting Infohabitants (EEII) project. The overall goal of the EEII project was to explore the use of ecological methods within the development of agent-based information systems. The project focused on studying agent systems from an ecological standpoint and properties of agent systems such as scalability [Marwala et al., 2001], openness [Abramov et al., 2001], stability [Chli et al.,

[1] Part of this paper is based on Stability of Multi-agent Systems, by M. Chli et al. which appeared in Proceedings of the 2003 IEEE International Conference on Systems, Man and Cybernetics. 2004 IEEE. Reprinted by permission of IEEE.

M. Chli and P. De Wilde, *Convergence and Knowledge Processing in Multi-Agent Systems*,
Advanced Information and Knowledge Processing,

2003] and adaptability [Simoes-Marques et al., 2003] were investigated. A concise review of the EEII project is given in Appendix A.

The objective work described in this chapter was to extend an earlier developed notion of stability [De Wilde et al., 1999a; Lee et al., 1998] from an agent context to an ecology of infohabitants context. In these works the system of agents was considered stable if the proportion of agents that executed tasks was stationary. This notion of stability has been refined and adapted so that it is suitable for multi-agent systems with a varying number of agents (ecosystems). The proposed definition for stability is inspired from Markov Systems . The notion of stationarity, which is related to the population dynamics theory [Weibull, 1995], is not relevant here.

For illustration and verification purposes the proposed definition was applied to simple games. It was then applied to more complex agent systems/ecosystems, specifically to two models (trading and load transportation) that were implemented for this purpose. Extensive experiments were conducted on these models, and quantitative measures were used to devise whether they were stable under a wide range of conditions.

The rest of the chapter is organised as follows. The following section reviews other approaches to stability and explains why they are not suitable for multi-agent systems. In Section 3.3 the proposed definition of stability of multi-agent systems is presented and its application is illustrated on multi-agent games. Section 3.4 details the models used for the investigation as well as the experiments performed and the results produced. Section 3.5 concludes the discussion.

3.2 Background

Multi-agent systems is a growing field mainly because of the recent development of the Internet as a means of circulating information, goods and services. Many researchers have contributed valuable work in the area in recent years. The global properties of multi-agent systems (e.g. openness, scalability) have been widely investigated. However, what is still missing is a clear notion of the stability of multi-agent systems.

The agents of a multi-agent system are computer programs in a distributed environment that execute tasks on behalf of their human owners. These tasks often involve decision making. Stability, understood intuitively as the property of a system that exhibits bounded behaviour, is perhaps the most desired feature in the systems we design. It is important for us to be able to predict the response of a multi-agent system to various environmental conditions prior to its actual deployment. This is why we believe that a clear mathematical definition of the concept will allow us to develop tools and methods for its analysis. This definition must differentiate the notion of stability in multi-agent systems from that of stationarity [De Wilde et al., 1999b]. A stationary variable has the property that the mean does not change over time [Weibull, 1995]. Stationarity can be defined in precise mathematical terms, but for our purpose we mean a flat looking series, without trend, a constant autocorrelation structure over time and no periodic fluctuations.

Computer scientists often talk [Balakrishnan et al., 1997; Thomas and Sycara, 1998] about stable or unstable systems without having a concrete and uniform definition of stability. For example, in Thomas and Sycara [1998] stability is referred to in a way similar to the game theoretic definition, specifically, focussing on how to get the agents to distribute themselves among resources such that none of them have incentives to switch. In game theory an equilibrium point is a set of strategies from all players wherein each strategy is the best reply to others. An equilibrium point is considered stable if no player has any incentive to deviate from this equilibrium. In some other computer science work stability of a system is interpreted differently. For instance, in Balakrishnan et al. [1997] two forms of stability are identified and analysed: spatial stability, which determines the variation in the observed parameter for an entity in comparison to its neighbours, and temporal stability, which determines the time scales over which various observed parameters are valid. In that work stability is approached from a stochastic point of view in terms of an equilibrium distribution.

On the other hand, control engineers have a very well-established definition [Khalil, 1995], which however, is not suitable for multi-agent systems. This is mainly because we model the behaviour of agents that have to make decisions which is different from changing the values of variables, performing arithmetical operations or differentiation and integration and is harder to automate. In addition to this, the stability notion of control theory does not allow for oscillations (e.g. sinusoidal oscillations) in the system, in this way favouring stationary systems.

The widely accepted notion of stability, which comes from the field of dynamical systems [Perko, 2001; Glance et al., 1991], is not appropriate for multi-agent systems either. It states that an equilibrium is stable if after a small perturbation the system returns to its equilibrium "voluntarily" (e.g. as a pendulum does). This definition is restricted for closed systems. Agents, as they act on behalf of humans, are not isolated from the real world. They constantly have to deal with new concepts and changing input.

Similarly the definition of stability in the context of population dynamics is not adequate [Weibull, 1995]. In evolutionary game theory the research concentrates on grouping agents into types and studying how populations of different types evolve by looking at the size of the population. A population is in an evolutionary stable state if its genetic composition is restored by selection after a disturbance. However, in multi-agent systems used to model complex phenomena the agents may not always fall into classes, as they are independent individuals.

We propose that considering agents as utility-maximising players who take decisions in a game, instead of computer programs, would be a more appropriate approach that concentrates on the actions of the agents rather than implementation issues. In addition, we want to account for systems that are as close to reality as possible. In order to do this we have to cater for systems with a varying number of agents, which is why we work with *ecosystems of agents*. In an ecosystem, new agents can appear and existing agents can vanish from the system.

In this work we propose a definition of stability for multi-agent systems based on the stationary distribution of a stochastic system. This is the only definition that takes into account the game nature of multi-agent systems, is valid for systems with varying number of agents and is fully supported by the mathematical framework of stochastic processes. We provide example systems to illustrate this.

3.3 Stability in Games

3.3.1 Stochastic Systems Primer

This section presents a brief and by no means exhaustive introduction to the theory of Markov chains, which underlies the approach taken in this work to defining stability. More comprehensive introductions to Markov chain theory and stochastic processes in general are available in Norris [1997] and Cox and Miller [1984].

The first part is an overview of the fundamentals of Markov chains reviewing the basic definitions and the markovian property. We consequently introduce the definition of stability of multi-agent systems.

3.3.1.1 Definition and Basic Properties of Discrete-Time Markov Chains

Let I be a countable set. Each $i \in I$ is called a *state* and I is called the *state-space*. We say that $\lambda = (\lambda_i : i \in I)$ is a *measure on* I if $0 \leq \lambda_i < \infty$ for all $i \in I$ and a *distribution* if additionally $\sum_{i \in I} \lambda_i = 1$.

If X is a random variable taking values in I and we have that $\lambda_i = Pr(X = i)$ then λ is *the distribution of* X. We say that a matrix $P = (p_{ij} : i, j \in I)$ is *stochastic* if every row $(p_{ij} : j \in I)$ is a distribution.

We will extend our familiar notions of matrix and vector multiplication to cover a general index set I that may potentially be infinite in size. Namely we will define the multiplication of a matrix by a measure as a new measure λP given by

$$(\lambda P)_i = \sum_{j \in I} \lambda_j p_{ij}. \tag{3.1}$$

We shall now describe the rules for a Markov chain by a definition in terms of the corresponding matrix P.

Definition 3.1. We say that $(X_n)_{n \geq 0}$ is a *Markov chain* with *initial distribution* $\lambda = (\lambda_i : i \in I)$ and *transition matrix* $P = (p_{ij} : i, j \in I)$ if:

1. $Pr(X_0 = i_0) = \lambda_{i_0}$ and
2. $Pr(X_{n+1} = i_{n+1} \mid X_0 = i_0, \ \ldots, \ X_n = i_n) = p_{i_n i_{n+1}}$.

We abbreviate these two conditions by saying that $(X_n)_{n \geq 0}$ is *Markov*(λ, P). The following theorem is an easy consequence of condition 2.

Theorem 3.1. *A discrete-time random process* $(X_n)_{n\geq 0}$ *is* $Markov(\lambda,P)$ *if and only if for all* N *and* $i_0,\ldots,i_N$ *we have*

$$Pr(X_0 = i_0,\ldots,X_N = i_N) = \lambda_{i_0} p_{i_0 i_1} \cdots p_{i_{N-1} i_N}. \tag{3.2}$$

This theorem depicts the structure of a Markov chain. Namely, it illustrates the relation with the stochastic matrix P and reveals the time-invariance property of a Markov chain.

Theorem 3.2. *Let* $(X_n)_{n\geq 0}$ *be* $Markov(\lambda,P)$. *Then for all* $n,m \geq 0$
(i) $Pr(X = j) = (\lambda P^n)_j$,
(ii) $Pr(X_n = j \mid X_0 = i) = Pr(X_{n+m} = j \mid X_m = i) = (P^n)_{ij}$.

We sometimes denote $(P^n)_{ij}$ as $p_{ij}^{(n)}$. So in light of Theorem 3.2 we define $p_{ij}^{(n)}$ as the n-step transition probability from i to j.

We now introduce the concept of an *invariant distribution*. We say that λ is invariant if

$$\lambda P = \lambda. \tag{3.3}$$

The following theorem links the existence of an invariant distribution which is an algebraic property of the matrix P with the probabilistic concept of an *equilibrium distribution*. It applies to a somewhat restricted class of chains, namely those with irreducible and aperiodic stochastic matrices. There is, however, a multitude of similar results for other types of chains for which we refer to Norris [1997] and Cox and Miller [1984]. This result is only provided as an indication of this family of theorems. An *irreducible* matrix P is one for which for all $i,j \in I$ and for sufficiently large n, $p_{ij}^{(n)} > 0$. It is *aperiodic* if for all states $i \in I$ we have $p_{ii}^{(n)} > 0$ for all sufficiently large n. We can now state:

Theorem 3.3. *Let* P *be irreducible and aperiodic, and suppose that* P *has an invariant distribution. Let* λ *be any distribution. Suppose that* $(X_n)_{n\geq 0}$ *is* $Markov(\lambda,P)$. *Then*

$$Pr(X_n = j) \to \pi_j \text{ as } n \to \infty \text{ for all } j \in I \tag{3.4}$$

and

$$p_{ij}^{(n)} \to \pi_j \text{ as } n \to \infty \text{ for all } i,j \in I. \tag{3.5}$$

3.3.2 *Definition of Stability*

We view the system as a countable set of states I with implicitly defined transitions P between them. At time n, the state of the system is the random variable X_n. The key assumption of this work is that $(X_n)_{n\geq 0}$ is $Markov(\lambda,P)$.

Definition 3.2. A system considered under those terms will be said to be **stable** when the state of the system converges into an equilibrium distribution. In other words, when

$$Pr(X_n = j) \to \pi_j \text{ as } n \to \infty \text{ for all } j \in I. \tag{3.6}$$

Our definition can also be stated by: *A Markov process $x_1, x_2, x_3, x_4, \ldots$ is **stable** if, the probability distribution of x_m becomes independent of the time index m for large m.*

Most Markov chains with a finite state space and positive transition probabilities are examples of stable systems because after a "burn-in" period they settle down on a stationary distribution.

Below, we provide example games to illustrate our definition of stability.

3.3.2.1 Dealing with a Varying Number of Agents

We require that the definition of stability we propose applies to systems with a varying number of agents. It is therefore important to be able to portray whether an agent is present in the system. The system state will be represented by a finite vector **X**, with dimensions large enough to cope with a reasonable amount of agents present in the system. The state vector will consist of one or more elements for each agent in addition to a number of elements to describe general properties of the system state that are not particular to any single agent. We model an agent as being "dead", i.e. not being present in the system, by setting the vector elements for that agent to some predefined value.

3.3.3 Example Games

A couple of the example "toy" games we dealt with are given below. We will explain how the proposed definition of stability applies to them and then investigate how more complicated games can be treated.

The method we follow to assess the stability of system is first to derive the transition matrix $\mathscr{P}$ of the process from the rules of the game. The transition matrix is used to predict whether the multi-agent system game under investigation has an equilibrium distribution, to conclude whether it is stable or not.

3.3.3.1 The 3-Player $+/-$ Game

The three players of this game, can each be in one of two states "$+$" or "$-$".

- If two players are in the same state the third player goes into that state.
- If all three players are in the same state, there is probability 0.9 that they will all remain in that state and probability 0.1 that they will all change state.

This game has eight states which are shown in Table 3.1.

Table 3.1 The states and transition matrix for the 3-player +/− game.

(a) The states

Id	State
1	− − −
2	− − +
3	− + −
4	+ − −
5	+ − +
6	− + +
7	+ + −
8	+ + +

(b) The transition matrix

State Id	1	2	3	4	5	6	7	8
1	0.9	0	0	0	0	0	0	0.1
2	1	0	0	0	0	0	0	0
3	1	0	0	0	0	0	0	0
4	1	0	0	0	0	0	0	0
5	0	0	0	0	0	0	0	1
6	0	0	0	0	0	0	0	1
7	0	0	0	0	0	0	0	1
8	0.1	0	0	0	0	0	0	0.9

The state transition probabilities can be derived from the rules of the game. For example, the probability of going to state 8 (+ + +) from state 1 (− − −) is 0.1, the probability of remaining in state 8 (+ + +) 0.9, the probability of going to state 8 from state 7 (+ + −) is 1, or the one of going to state 4 (+ − −) from state 6 (− + +) is 0 and so on. Thus the 8 × 8 state transition matrix shown in Table 3.1 is constructed.

The transition matrix is often referred to as $\mathscr{P}$. Once we have this we can check whether an equilibrium distribution exists by solving the following simultaneous equations for π:

$$\pi\mathscr{P} = \pi, \tag{3.7}$$

$$\sum \pi_i = 1. \tag{3.8}$$

If *one or more solutions* exist then the system is stable. In the case where only one solution exists, this solution will be the equilibrium distribution into which the system will converge. If more than one solution exist, this means that the system can converge into one of many stable solutions/distributions of states. It is uncertain in which one the system will fall, but as soon as it does it then remains in that one forever and thus it is considered stable. For the 3-player +/− game described in this section, one solution exists, yielding an equilibrium distribution $\pi = [0.5\ 0\ 0\ 0\ 0\ 0\ 0\ 0.5]^T$.

With the aid of this example it has been illustrated how state transition probabilities can be derived from the rules of a game, which was formulated from a multi-agent system scenario. These transition probabilities allow us to predict whether the system will eventually reach an equilibrium distribution, in other words whether the system is stable.

3.3.3.2 The Simple Random Walk

An example of an unstable system is the simple random walk. Starting in state 0 at each time-step, if we are in state K, we go either to state $K+1$ or to state $K-1$ with equal probabilities. It can be shown that the distribution of the state X_m, where m is the time, is centered on zero but its variance is proportional to $\sqrt{m}$ which means that the p.d.f. is not independent of m, making the system unstable.

3.3.3.3 A Coin Game

In another slightly more complicated game, N players toss a coin in pairs.

- The winner gives to the loser a unit of its wealth.
- Players are destroyed if their wealth is below 1 and can generate another player if their wealth is more than 4.
- The sum of all players' wealth should be 6.

When this game is played with three players its state space consists of 19 states. A state of the game is of the form $W_1 W_2 \cdots W_n$, where W_i is the wealth of the ith player. A "–" denotes an agent that is no longer in the game because its wealth has reached 0. The state space is shown in Table 3.2(a).

The transition matrix $\mathscr{P}$ of the process is derived from the rules of the game. This is a 19×19 matrix which shows the probability of reaching state j given that the current state is i. The transition matrix is used to predict whether this multi-agent system game has an equilibrium distribution, i.e. whether it is stable. Below is an example of how the transition matrix is formulated.

Suppose the system is in state 12 (213), i.e. $Agent_1$ has $wealth = 2$, $Agent_2$ has $wealth = 1$ and $Agent_3$ has $wealth = 3$. There are three possibilities, each with equal probability of taking place: either $Agent_1$ plays with $Agent_2$, or $Agent_2$ with $Agent_3$, or $Agent_1$ with $Agent_3$.

Each agent has equal probability of winning the game, i.e. taking one of its opponent's wealth.

We find from the probability tree, shown in Figure 3.1, that each of the six final possibilities have probability $\frac{1}{2} \times \frac{1}{3} = \frac{1}{6}$ to take place. Therefore, when the system is in state 12 has probability $\frac{1}{6}$ to go to each of the states 2, 8, 10, 14, 16 and 17 in the next step and probability 0 to go into any other state.

We repeat this for each one of the 19 states in the game. The 19×19 state transition matrix is shown in Table 3.2(b). Solving the system of simultaneous equations given by we equations (3.7) and (3.8) we show that the game has an equilibrium distribution

$$\pi = [0.0625\ 0.0625\ 0.0625\ 0.0521\ 0.0521\ 0.0521\ 0.0521\ 0.0521\ 0.0521\ 0.0312\ 0.0313\ 0.0312\ 0.0313\ 0.0312\ 0.0312\ 0.0312\ 0.0937\ 0.0938\ 0.0937]^T.$$

(a) States

Id	State
1	33-
2	3-3
3	-33
4	42-
5	4-2
6	-42
7	24-
8	2-4
9	-24
10	123
11	132
12	213
13	231
14	312
15	321
16	222
17	114
18	141
19	411

(b) Transition matrix

State Id	1	2	3	4	5	6	7	8	9	10	11	12	13	14	15	16	17	18	19
1	0	0	0	$\frac{1}{2}$	0	0	$\frac{1}{2}$	0	0	0	0	0	0	0	0	0	0	0	0
2	0	0	0	0	$\frac{1}{2}$	0	0	$\frac{1}{2}$	0	0	0	0	0	0	0	0	0	0	0
3	0	0	0	0	0	$\frac{1}{2}$	0	0	$\frac{1}{2}$	0	0	0	0	0	0	0	0	0	0
4	$\frac{1}{2}$	0	0	0	0	0	0	0	0	0	0	0	0	0	0	0	0	0	$\frac{1}{2}$
5	0	$\frac{1}{2}$	0	0	0	0	0	0	0	0	0	0	0	0	0	0	0	0	$\frac{1}{2}$
6	0	0	$\frac{1}{2}$	0	0	0	0	0	0	0	0	0	0	0	0	0	0	$\frac{1}{2}$	0
7	$\frac{1}{2}$	0	0	0	0	0	0	0	0	0	0	0	0	0	0	0	0	$\frac{1}{2}$	0
8	0	$\frac{1}{2}$	0	0	0	0	0	0	0	0	0	0	0	0	0	0	$\frac{1}{2}$	0	0
9	0	0	$\frac{1}{2}$	0	0	0	0	0	0	0	0	0	0	0	0	0	$\frac{1}{2}$	0	0
10	0	0	$\frac{1}{6}$	0	0	0	0	0	$\frac{1}{6}$	0	$\frac{1}{6}$	$\frac{1}{6}$	0	0	0	$\frac{1}{6}$	$\frac{1}{6}$	0	0
11	0	0	$\frac{1}{6}$	0	0	$\frac{1}{6}$	0	0	0	$\frac{1}{6}$	0	0	$\frac{1}{6}$	0	0	$\frac{1}{6}$	0	$\frac{1}{6}$	0
12	0	$\frac{1}{6}$	0	0	0	0	0	$\frac{1}{6}$	0	$\frac{1}{6}$	0	0	0	$\frac{1}{6}$	0	$\frac{1}{6}$	$\frac{1}{6}$	0	0
13	$\frac{1}{6}$	0	0	0	0	0	$\frac{1}{6}$	0	0	0	$\frac{1}{6}$	0	0	0	$\frac{1}{6}$	$\frac{1}{6}$	0	$\frac{1}{6}$	0
14	0	$\frac{1}{6}$	0	0	$\frac{1}{6}$	0	0	0	0	0	0	$\frac{1}{6}$	0	0	$\frac{1}{6}$	$\frac{1}{6}$	0	0	$\frac{1}{6}$
15	$\frac{1}{6}$	0	0	$\frac{1}{6}$	0	0	0	0	0	0	0	0	$\frac{1}{6}$	$\frac{1}{6}$	0	$\frac{1}{6}$	0	0	$\frac{1}{6}$
16	0	0	0	0	0	0	0	0	0	$\frac{1}{6}$	$\frac{1}{6}$	$\frac{1}{6}$	$\frac{1}{6}$	$\frac{1}{6}$	$\frac{1}{6}$	0	0	0	0
17	0	0	0	0	0	0	0	$\frac{1}{6}$	$\frac{1}{6}$	$\frac{1}{6}$	0	$\frac{1}{6}$	0	0	0	0	$\frac{1}{3}$	0	0
18	0	0	0	0	0	$\frac{1}{6}$	$\frac{1}{6}$	0	0	0	$\frac{1}{6}$	0	$\frac{1}{6}$	0	0	0	0	$\frac{1}{3}$	0
19	0	0	0	$\frac{1}{6}$	$\frac{1}{6}$	0	0	0	0	0	0	0	0	$\frac{1}{6}$	$\frac{1}{6}$	0	0	0	$\frac{1}{3}$

Table 3.2: The states and transition matrix for a coin game for three players.

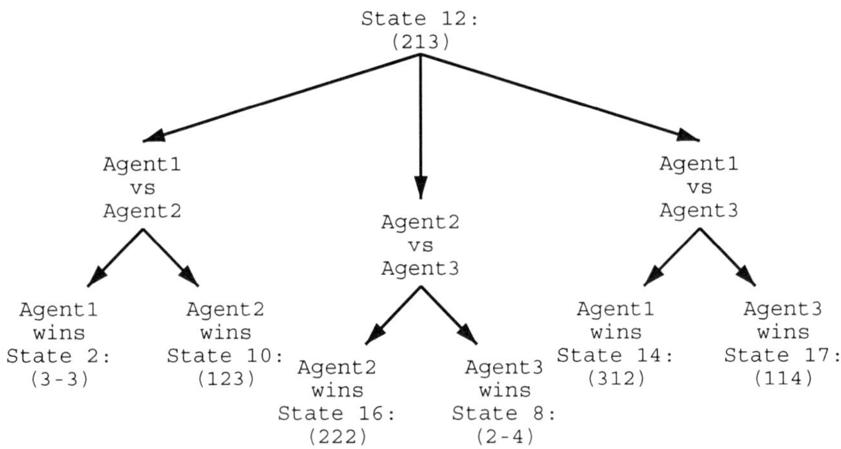

Fig. 3.1: Coin game for three players: Probability tree for state 12.

Note that even for a simple game with such a small number of players the number of states is significant. The state space size rises dramatically with the number of players. For example, the same game with four players instead of three has a state space of around 70 states.

At least theoretically, one could evaluate the transition probabilities of the multi-agent system under investigation and find its equilibrium distribution, if it exists. In practice, due to the state space explosion problem in more realistic multi-agent systems, this approach is not viable. We therefore propose statistical analysis as a means of testing for the probabilistic behaviour of more complex systems.

Multi-agent systems are often used to analyse problems such as trading in a stock exchange or transportation problems. This type of situation can be formulated as games and their stability can be analysed in the way proposed in this section, or with statistical analysis. In the next section, it is shown how a trading scenario, a loads transportation scenario and a virus-spreading scenario are formulated as games and their stability is comprehensively analysed. The results obtained from extensive simulations and statistical analysis are presented.

3.4 Experiments

In order to test the proposed definition of stability we have experimented with various models of multi-agent systems that have been implemented in Java. These models, being closer to real-life systems, are governed by significantly larger and more complicated rule sets than the example games described in the previous section. It is therefore not practical to go about determining the state space and the transition matrix of each one of them. Instead, several indicative metrics are examined to assess whether they reach a stationary distribution after the system has been left to run for a while. With the aim of exploring the whole range of the parameter space for our models an automated mechanism to carry out a large number of experiments was set up. In the sections that follow, the models are described and the most indicative experiments are shown.

3.4.1 Trading Simulation Model

3.4.1.1 Initial Setting

There is an array of N different resources in this model. There are M traders each with its wealth, its discounting percentage and its available resources. All traders are endowed with the same amount of wealth; the amount of each resource given to each trader is randomly calculated. A discounting percentage is used when the trader recalculates its prices. Its function is explained in more detail in "Pricing of the Resources" below. Finally, the scenario involves tasks that are generated by agents. A task requires some resources and produces some others, loosely modelling generic economic activity.

At each clock-tick every trader with its turn issues a task and advertises it amongst the other traders. Each task carries a random combination of required and produced resources.

Every trader in the multi-agent system gives an offer for the task (provided that they possess the required resources). The cheapest offer is selected. If the issuer cannot pay for any offer then the task is not executed. Otherwise, it selects an offer and the task is executed. The required resources are subtracted from the task executor's set of resources, the produced resources are added to the issuer's set of resources and the issuer pays to the executor an amount of money equal to the price for executing the task. The flow of the algorithm is illustrated in Figure 3.2 below.

Pricing of the Resources

The price each trader sets for each resource is different. After the execution of a task all the traders that gave offers for that task recalculate their prices. Only the prices of the task's required resources are altered on recalculation; the rest remain as they were. There are three ways in which recalculation occurs:

1. The trader whose offer was accepted increases the prices of the required resources of the task as follows:
   ```
   resourcePrices[i] +=
   this.resourcePrices[i] * discountingFactor;
   ```
2. The traders whose offers were not accepted decrease the prices of the required resources of the task as follows:
   ```
   resourcePrices[i] -=
   (1 - selectedPrice/myPrice)*
   this.resourcePrices[i] * discountingFactor;
   ```
3. In case no offer was accepted all traders that gave offers for the task decrease the prices of the required resources of the task as follows:
   ```
   resourcePrices[i] -=
   (1 - maxPricePayable/myPrice) *
   discountingFactor * this.resourcePrices[i];
   ```

Generation and Destruction of Agents

When a trader is sufficiently rich, i.e. its wealth exceeds a certain threshold, it generates a new trader to which it gives half its wealth. Also, the parent trader endows the child trader with half of its resources. The new trader inherits its generator's discounting factor. Conversely, when a trader's wealth goes below zero, then it is destroyed and removed from the system.

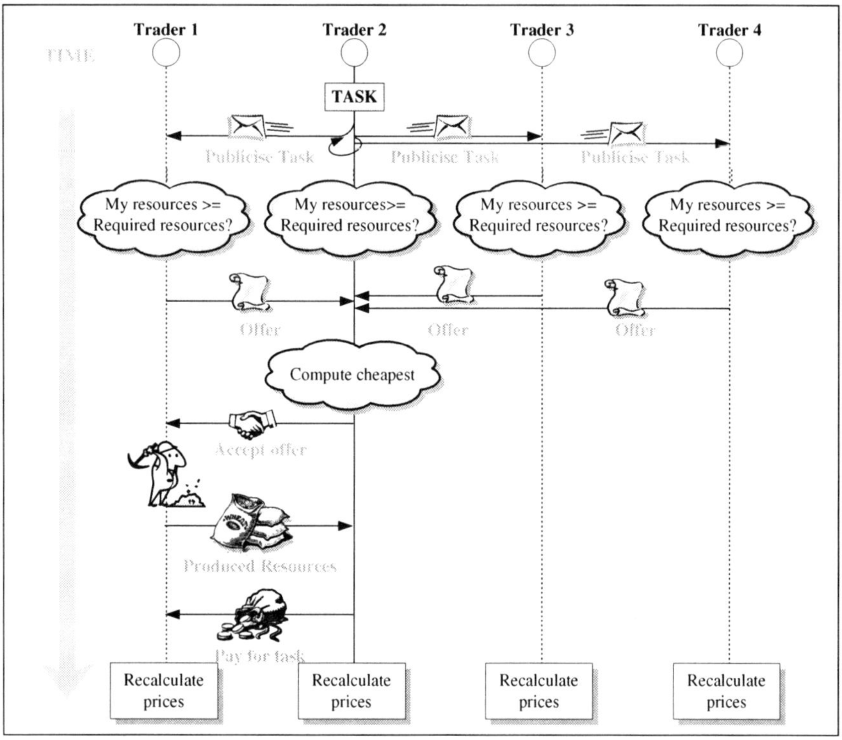

Fig. 3.2: Flow diagram of the trader's simulation model algorithm during one time tick.

3.4.1.2 Stability Conditions

In order to consider the system stable, its state must reach a stationary distribution when it has been left to run for a sufficient amount of time. A strong indication of stability would be that system metrics such as

- the proportion of traders that execute tasks,
- the number of traders,
- the prices of the resources and
- the wealth per trader

would all reach stationary distributions when the system has been left to run for a while.

3.4.1.3 Experiments

In order to explore the entire parameter space, an extensive and exhaustive set of experiments was performed, for as many combinations of initial parameter values and perturbation sizes as possible. The parameter values that were varied for this model were the following: the order of magnitude of the prices, resources and discounting factor as well as the initial, generation and destruction wealth of the traders. A selection of the most interesting experiments performed is presented below.

An Unstable System

An experiment was performed with 50 traders and 10 different types of resources. The simulation was left to run for 10,000 rounds. Each trader was endowed with 10^6 monetary units and a random amount of each of the 10 resources. The amount from each resource it gets was of the order of 1000 (calculated randomly). The resources' prices were initially of the order of 100 monetary units (calculated randomly). The discounting factor's initial order of magnitude was 10^{-3}. A trader could generate a new trader if its wealth exceeded 150,000 monetary units and it was removed from the game if its wealth went below 0. The parameters of this experiment are shown in detail in Table 3.3.

The graphs in Figure 3.3 show how the number of traders, the price of one of the resources, the wealth per trader and the tasks executed per trader vary during the course of the simulation. These are all metrics that we observe to assess whether the system is stable. Statistical analysis was used to evaluate the stability of the system. The aim is to show whether the metrics mentioned above reach a stationary distribution, after the system has been left to run for some time. For this to happen the values for each metric must fall into a distribution. Hypothesis testing was used to establish whether this is the case. Some concepts from statistical hypothesis testing are explained in Appendix B.

Also in Appendix B hypothesis testing is performed on two samples of the **tasks executed per trader** metric. The test for the means succeeded. However, the test for the variances did not succeed. This means that there is no significant evidence that the two samples come from the same distribution. Therefore, the system is *not stable* when run with the initial conditions mentioned above.

Similar tests were performed for all of the metrics, which show analogous results. In general, in order to accept that a system is stable it is required that all of the metrics exhibit stability.

A Stable System

If the experiment described above is rerun with the same initial parameters, only changing the discounting factor's order of magnitude from 10^{-3} to 10^{-13} and the prices' order of magnitude from 10^3 to 10^4, the system's behaviour is significantly different. The parameters of this experiment are given in detail in Table 3.4.

It is evident from the graphs in Figure 3.4 that the system is more stable than in the previous experiment. Hypothesis tests for each metric were performed as described above. For all of them there was significant evidence in the 5% significance

Simulation Parameters	
Number of runs	10,000
Number of traders	50
Number of resources	10
Discounting factor's order of magnitude	0.001
Initial wealth	100,000
Prices' order of magnitude	100
Resources' order of magnitude	1000
Generation wealth	150,000
Destruction wealth	0

Table 3.3: Trading Scenario, an unstable system: Experiment parameters.

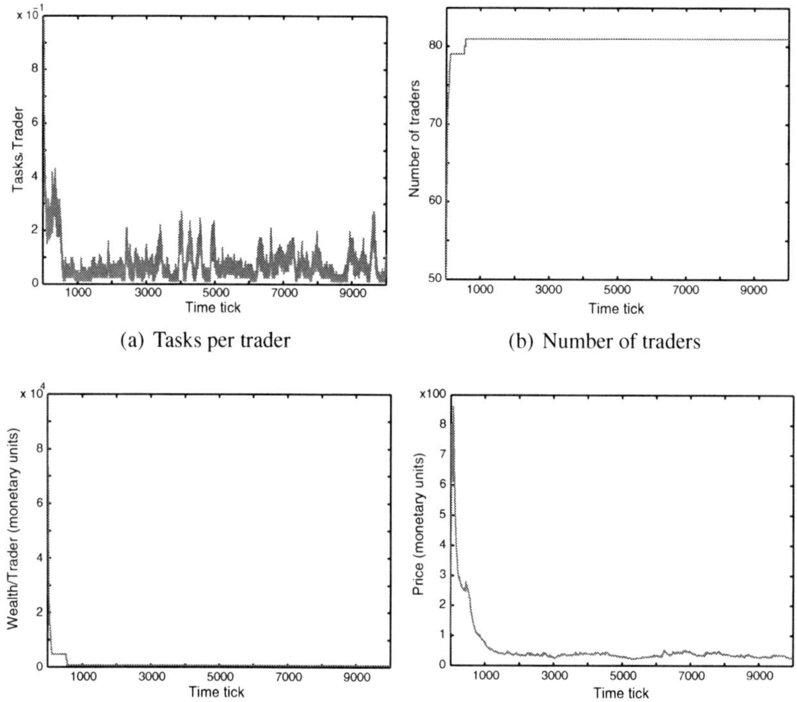

(a) Tasks per trader

(b) Number of traders

(c) Wealth per trader

(d) Price of resource 5

Fig. 3.3: Trading scenario, an unstable system. To conclude that a system is stable it is required that all the metrics reach a stationary distribution after the system has been left to run for a while. It is evident from the graph that at least the tasks per trader metric does not fall in a stationary distribution. This observation is justified by statistical analysis.

Simulation Parameters	
Number of runs	10,000
Number of traders	50
Number of resources	10
Discounting factor's order of magnitude	10^{-13}
Initial wealth	100,000
Prices' order of magnitude	100
Resources' order of magnitude	10,000
Generation wealth	150,000
Destruction wealth	0

Table 3.4: Trading scenario, a stable system: Experiment parameters.

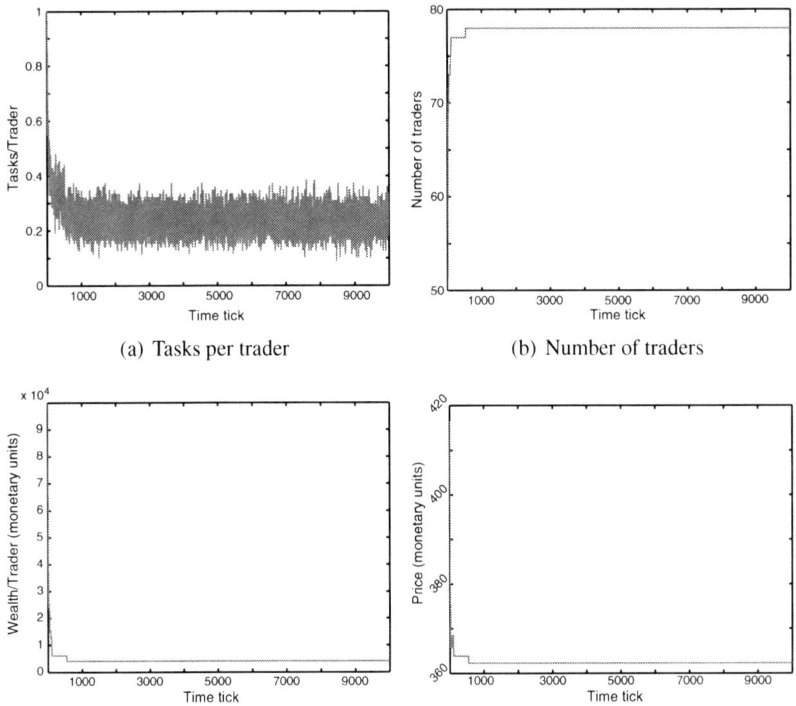

(a) Tasks per trader

(b) Number of traders

(c) Wealth per trader

(d) Price of resource 5

Fig. 3.4: Trading scenario, a stable system. It is evident from the graphs of the characteristic metrics that the system is stable. This observation is justified by the statistical analysis.

level the metrics reached a stationary distribution. Therefore the system is stable when run under those specific initial conditions.

Table 3.5 contains the details of the tests performed for the metric tasks per trader.

(a) t-test for equality of means

Significance Level 0.05		
	Sample 1	**Sample 2**
Mean	0.237453	0.237049
Variance	0.001678	0.001752
Observations	4001	4001
DoF	8000	
t stat	0.435980226	
$P(T <= t)$ 2-tail	0.662862846	
t crit. 2-tail	±1.960261216	
t stat in crit.region	No	
Accept H_0	Yes	

(b) F-test for equality of variance

Significance Level	0.05	
	Sample 1	**Sample 2**
Mean	0.237453	0.237049
Variance	0.001678	0.001752
Observations	4001	4001
DoF	4001	4001
F stat	0.957581019	
F crit. 2-tail lower	0.939893954	
F crit. 2-tail upper	1.063949817	
F stat in crit.region	No	
Accept H_0	Yes	

Table 3.5: Trading scenario, a stable system.: Statistical analysis on the metric tasks per trader. Because the original hypotheses are both accepted we may conclude that both samples come from the same distribution.

Introducing Perturbations

It is interesting to observe the response in the behaviour of the system when its normal running is disturbed by introducing a perturbation. An experiment was set up with the same initial conditions as the first one described (which was unstable). At time tick 4000, a shock was injected into the system; the prices of all the resources of each trader were increased arbitrarily by 500%. The simulation was allowed to run until time tick 10,000. Full details on the parameters of this experiment are shown in Table 3.6. The graphs for this experiment are shown in Figure 3.5.

While the wealth per trader or the number of traders are not significantly affected by the shock introduced, we note that the tasks per trader ratio seems to follow a much more stable trend after recovering from the shock at time tick 4000. Statistical analysis confirms that at the 5% significance level there is significant evidence that the values for the metric tasks executed per trader after the shock come from a common distribution. On the other hand, the price of resource 5 progresses into a transient recovery phase after the perturbation; however, it does not become stable. The perturbation causes the prices to rise; for example, for resource 5 the mean price before the shock was 16.8, whereas after the shock it fluctuates around 21.1.

In another experiment, the system was started with the same initial parameters as the stable system shown above. Full details on the parameters of this experiment are shown in Table 3.7. At time tick 4000 the following shock was injected into the system. The wealth of each trader was increased by 60%. Figure 3.6 shows that the effect of the perturbation was significant. It was shown using hypothesis testing that after the recovery from the perturbation the values of each of the metrics fell into a common distribution. In general, the system after the substantial effect of the perturbation has elapsed went into a stable phase.

Simulation Parameters		**Perturbation Parameters**	
Number of runs	10,000	Start shock at tick	4000
Number of traders	50	Shock duration (ticks)	1
Number of resources	10	Parameter to shock	Prices
Discounting factor's order of magnitude	0.001	Shock magnitude (%)	500
Initial wealth	10,0000		
Prices' order of magnitude	100		
Resources' order of magnitude	1000		
Generation wealth	150,000		
Destruction wealth	0		

Table 3.6: Trading scenario, perturbation 1: Experiment parameters.

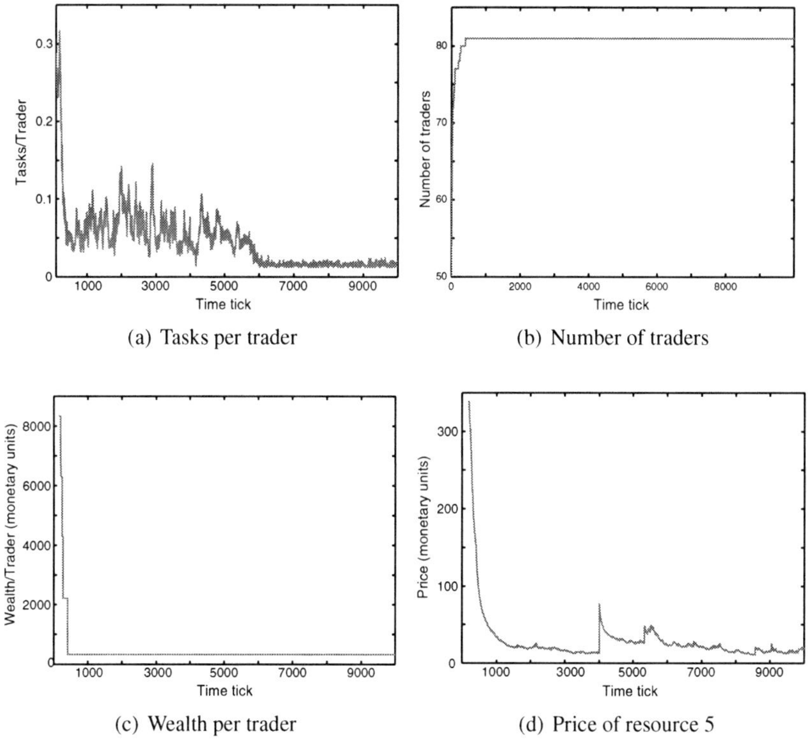

(a) Tasks per trader

(b) Number of traders

(c) Wealth per trader

(d) Price of resource 5

Fig. 3.5: Trading scenario, perturbation 1. The shock at round 4000 does not affect the wealth per trader and the number of traders. Statistical analysis shows that the tasks per trader ratio falls into a distribution after a recovery phase. On the other hand, the price of resource 5 is considerably affected but is not stabilised.

Simulation Parameters		Perturbation Parameters	
Number of runs	10,000	Start shock at tick	4000
Number of traders	50	Shock duration (ticks)	1
Number of resources	10	Parameter to shock	Wealth
Discounting factor's order of magnitude	10^{-13}	Shock magnitude (%)	-60
Initial Wealth	100,000		
Prices' order of magnitude	100		
Resources' order of magnitude	10,000		
Generation wealth	150,000		
Destruction wealth	0		

Table 3.7: Trading scenario, perturbation 2: Experiment parameters.

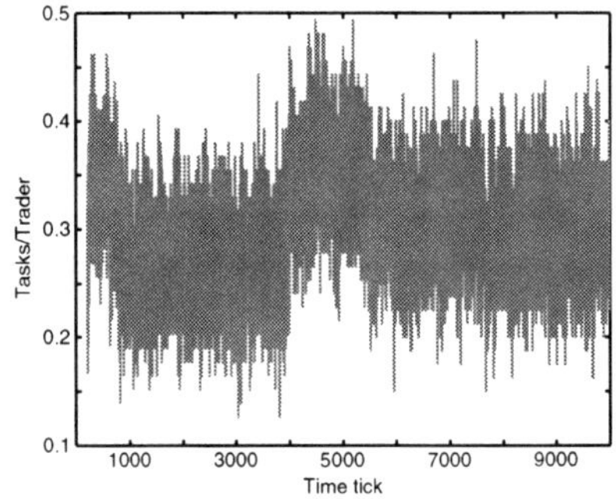

(a) Tasks per trader

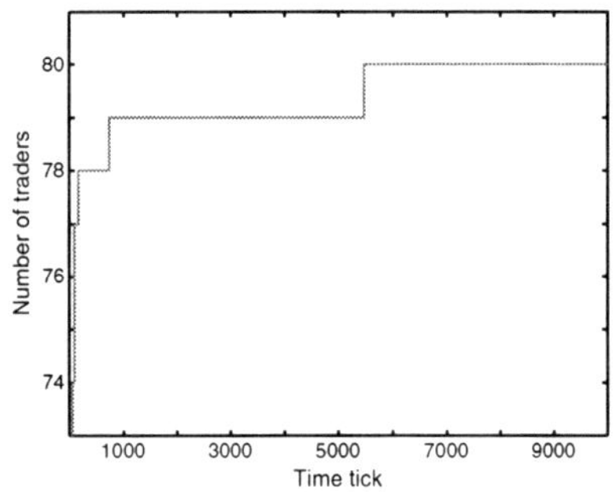

(b) Number of traders

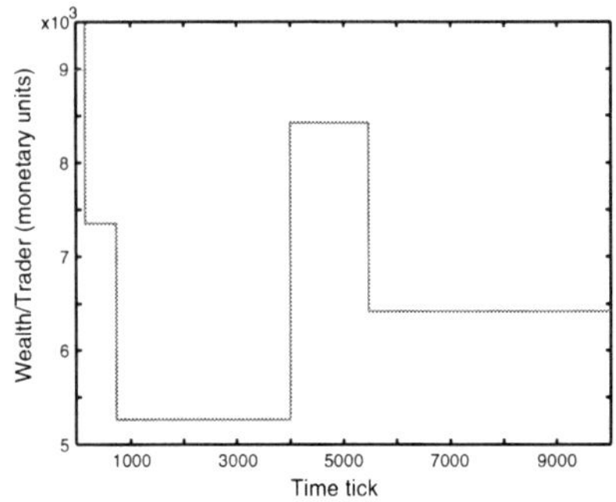

(c) Wealth per trader

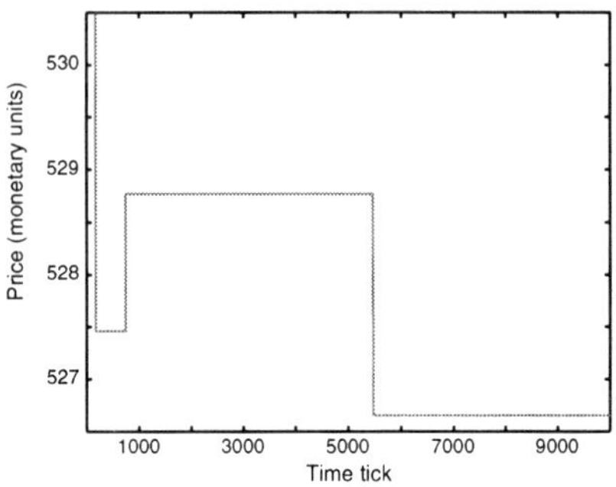

(d) Price of resource 5

Fig. 3.6: Trading scenario, perturbation 2. The perturbation introduced at round 4000 has a considerable effect on the metrics shown. Statistical analysis shows that after the system recovers it becomes stable.

3.4.2 Load Transportation Model

3.4.2.1 Initial Setting

We define a number N of cities and a map of interconnections between these cities. Some cities are directly connected to each other while some others are not. The distance between two cities that are directly connected takes one day to travel. For example, if the cities A and B are connected via city C (i.e.: A $\rightarrow$ C $\rightarrow$ B), it takes two days to travel from A to B. The simulation operates with a granularity of one day.

A number of M lorries travel around the map, each of them being at a city at any particular day. Up to M lorries can be at a given city at any time. Loads are generated at a certain rate at each city and have a specific city as a destination, e.g. a load can be generated at city A and be destined for city B. Each lorry can carry up to K loads, from which it earns rewards. Rewards are added to the lorry's wealth after the delivery of the load.

Load Allocation and Distribution

The loads located at a city are allocated to the lorries present at that city randomly. A lorry has an algorithm which decides the route it will follow based on information such as the city it currently is in and the destinations of the loads it carries.

Taxation

Every D number of days the lorries are required to reduce their wealth by a certain amount. This is a form of taxation.

Generation and Destruction of Agents

When a lorry's wealth exceeds a certain threshold, it generates a new lorry at the city it currently is and endows it with half its wealth. When a lorry's wealth is below a certain threshold then it is destroyed.

Example of the Model

Here we show how the model is applied in a scenario with two lorries that start travelling in a network of ten cities. The map of interconnections used in this example is shown in Figure 3.7(a).

The maximum capacity of a lorry is three loads. Lorries start with wealth equal to five and pay tax equal to one monetary unit every five days. A lorry is destroyed

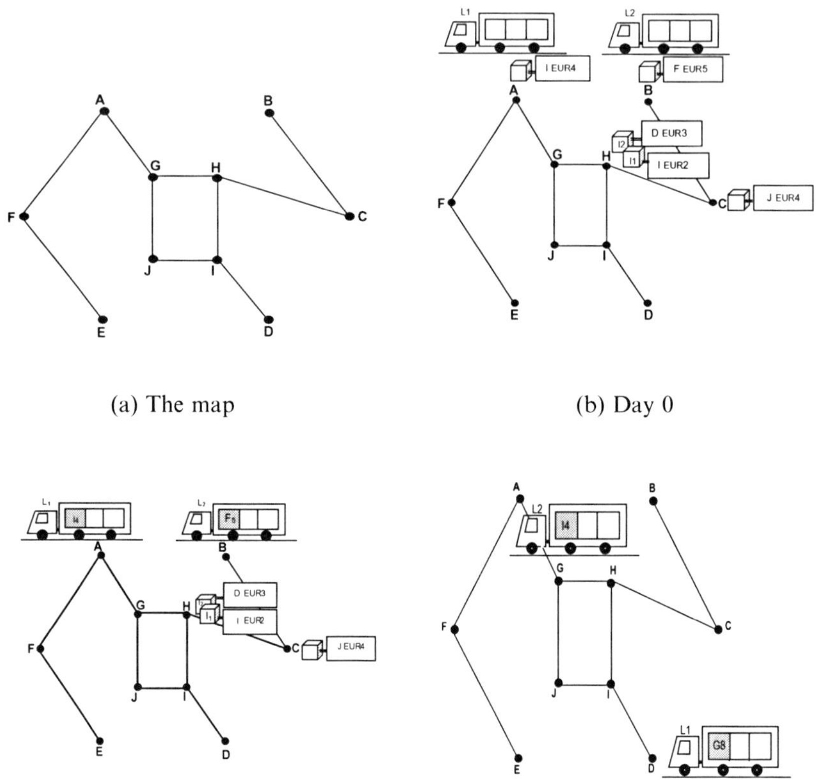

(a) The map (b) Day 0

(c) Day 1 (d) Day 8

Fig. 3.7: Load transportation: An example of the model in a setting with ten interconnected cities and two lorries.

when its wealth goes to zero. A lorry can generate a new entity when its wealth is greater or equal to 10, by keeping half its wealth and giving the other half to the new lorry. A description of the running of the simulation is given below, which also is depicted in Table 3.8.

On Day 0, $Lorry_1$ (L_1) starts at city A and $Lorry_2$ (L_2) starts at city B. A load with $X = 4$ (the potential reward) destined for city I waits in A and a load with $X = 5$ destined for city F waits in B. Also, there is a load in city C destined for J with $X = 4$ as well as two loads in city H; the first destined for D with $X = 3$ and the second for I with $X = 2$. This situation is illustrated in Figure 3.7(b).

On Day 1, L_1 takes the load that was waiting in A and L_2 takes the load that was waiting in city B. At this moment, each lorry has enough space to carry two more loads as shown in Figure 3.7(c). Each lorry plans the route it will follow in order

Day	Lorry1 (L1)					Lorry2 (L2)				
	Loc	**Action**	**W**	**Reward X**	**Route**	**Loc**	**Action**	**W**	**Reward X**	**Route**
1	A	Takes load for I	5	4 for l_I	GHI	B	Takes load for F	5	5 for l_F	CHGAF
2	G	No load	5	3 for l_I	HI	C	Takes load for J	5	4 for l_F + 4 for l_J	HGAF
3	H	Queue for Loads: L_2, L_1 Takes load for D	5	2 for l_I + 3 for l_D		H	Queue for Loads: L_2, L_1 Takes load for I	5	3 for l_F + 3 for l_J+ 2 for l_I	GAF
4	I	Unload	6	2 for l_D		G	No load	5	2 for l_F + 2 for l_J + 1 for l_I	AF
5	D	Unload	7	0		A	No load	5	1 for l_F + 1 for l_J + 0 for l_I	F
TAX			6	0				4		
6	D	Waits	6	0		F	Unload	4	0 for l_J - 1 for l_I	AGJ
7	D	New load for G at D	6	0		A	No load	4	-1 for l_J - 2 for l_I	GJ
8	D	Takes load for G	6	8 for l_G	IJG	G	No load	4	-2 for l_J - 3 for l_I	J
9	I	No load	6	7 for l_G	JG	J	Unload	1	-4 for l_I	I
10	J	No load	6	6 for l_G	G	I	Unload	-4		
TAX			5					-5		
						Lorry destruction				
11	G	Unload	10							
New lorry generation at G with wealth 5										
			5							

Table 3.8: Load transportation scenario: example run.

to reach the destination of the oldest load it carries while following the shortest possible path. Details are shown in Table 3.8. On Day 3, both L_1 and L_2 are in city H. In the same city there are two loads, l_1 and l_2, waiting for transportation. As the two lorries have space available they are interested in taking the loads. The load allocation is done as follows. The two lorries are arranged in a queue randomly. This is L_2, L_1. This means that L_2 is offered the chance to choose the load it prefers first and L_1 second.

On Day 4, L_1 delivers in I the load it took in A and is rewarded 1 monetary unit. It then plans the route it has to follow in order to deliver the other load it carries. On Day 5 it delivers this load in D. Day 5 is also Tax Day so 1 monetary unit is deducted from each lorry's wealth.

L_1 remains in city D until a load is generated there for transportation. On Day 10, L_2's wealth is -5. It is immediately destroyed. However, on Day 11 L_1's wealth reaches 10, so it generates a new lorry. L_1's wealth as well as that of the new lorry will be 5 monetary units.

3.4.2.2 Stability Conditions

According to the proposed definition of stability, the system will be considered stable if its metrics show evidence that they reach a distribution after some time the system has been running. The metrics observed to decide on whether the system is stable are:

- the proportion of lorries that carry loads,
- the proportion of loads that have been carried,
- the loads carried per lorry,
- the wealth per lorry, and
- the number of lorries.

3.4.2.3 Experiments

An Unstable System

An experiment was carried out with 100 lorries and a network of 500 cities connected to one another with a connectivity percentage of 40%. The simulation was left to run for 20,000 rounds. Each lorry was endowed with 10 monetary units. The maximum capacity of a lorry is three loads. A lorry can generate a new agent if its wealth exceeds 50 and it is destroyed if its wealth goes below 0 monetary units. Tax equal to 2 monetary units is deducted every five rounds. A new load is generated in each city every three rounds. The maximum number of loads that can be waiting in a city at any given time is five. The full list of the experiment parameters is given in Table 3.9.

The graphs in Figure 3.8 show how the characteristic metrics vary with time. Statistical analysis, comparing the last 4000 rounds to the 4000 rounds before those,

Simulation Parameters	
Number of runs	20,000
Number of lorries	100
Number of cities	500
Connectivity	40%
Tax period	5 days
Tax amount	2
Initial wealth	10
Generation wealth	50
Destruction wealth	0
Load generation frequency	1 load every 3 days
Load reward	3
Max loads in city	5

Table 3.9: Load transportation scenario, an unstable system: Experiment parameters.

was performed in order to decide on the system's stability. Hypothesis testing at the 5% significance level showed that after the system has been left to run for some time, most of the metrics mentioned above reach a stationary distribution.

This is not the case for the metric number of lorries. The statistical tests reveal there is significant evidence that the values for this metric do not fall into a specific distribution and for that reason we shall consider the system unstable. The details of these tests are shown in Table 3.10.

(a) t-test for equality of means

Significance Level	0.05	
	Sample 1	**Sample 2**
Mean	410.7813	408.9997
Variance	129.4789	127.6887
Observations	4001	4001
DoF	8000	
t stat	7.027085512	
$P(T \leq t)$ 2-tail	2.28214^{-12}	
t crit. 2-tail	± 1.960261216	
t stat in crit.region	Yes	
Accept H_0	No	

(b) F-test for equality of variance

Significance Level	0.05	
	Sample 1	**Sample 2**
Mean	410.7813	408.9997
Variance	129.4789	127.6887
Observations	4001	4001
DoF	4001	4001
F stat	1.01401972	
F crit. 2-tail lower	0.939893954	
F crit. 2-tail upper	1.063949817	
F stat in crit.region	No	
Accept H_0	Yes	

Table 3.10: Load transportation scenario, an unstable system: Statistical analysis on the metric "number of lorries". As the original hypotheses for the equality of means and variances are not both accepted, we conclude that the two samples do not come from the same distribution.

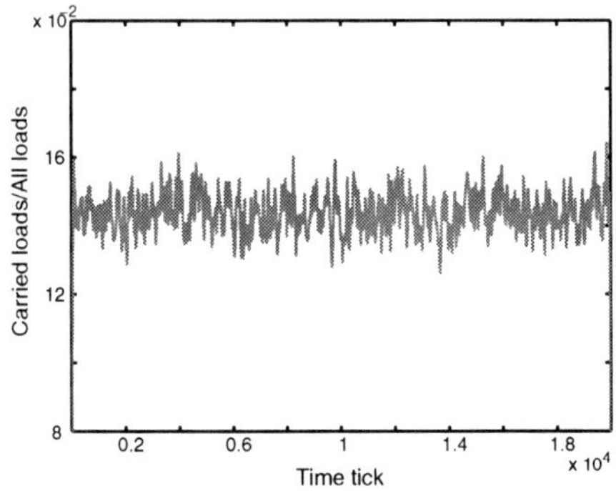

(a) Carried loads ratio

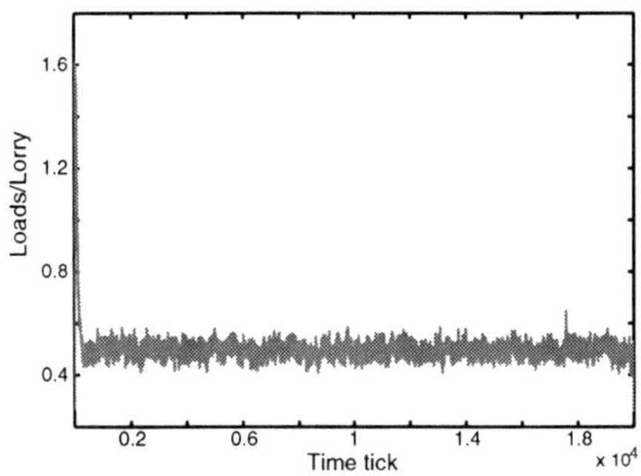

(b) Loads per lorry

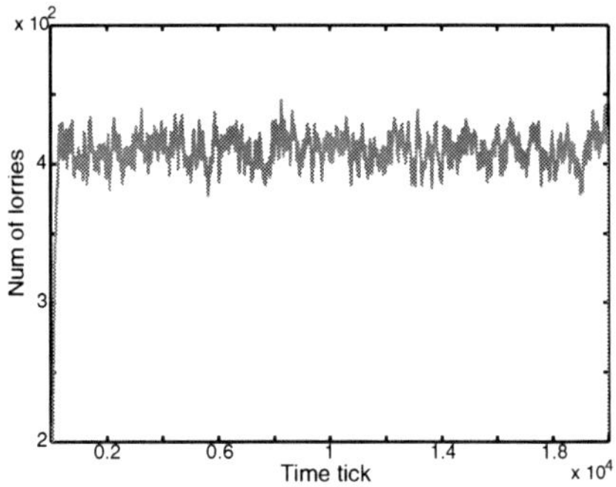

(c) Number of lorries

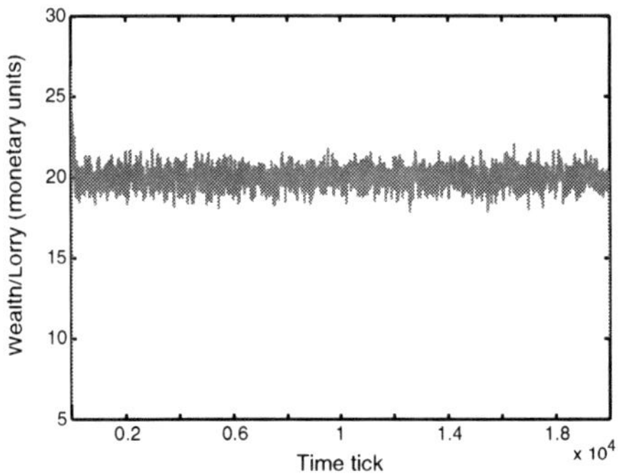

(d) Wealth per lorry

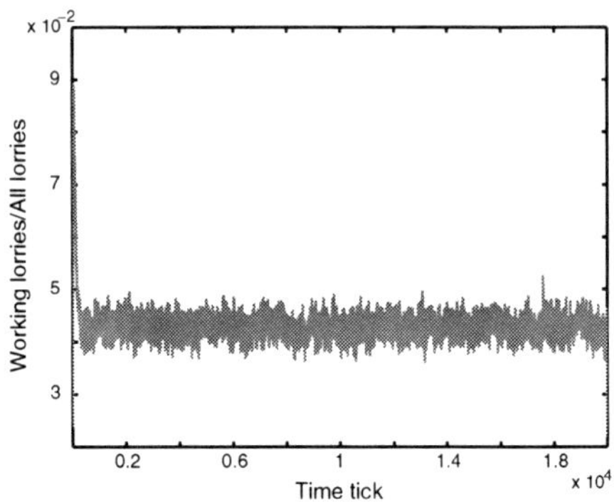

(e) Working lorries ratio

Fig. 3.8: Load transportation scenario, an unstable system. All the metrics except "number of lorries" reach a stationary distribution. The system is therefore considered unstable.

A Stable System

The previous experiment was rerun keeping the initial conditions the same with the exception of the intercity connectivity percentage. It was raised from 40% to 90%. This way, the map was more homogeneous.

Full details about the parameters of this experiment are shown in Table 3.11. The results obtained are shown in the graphs in Figure 3.9.

Simulation Parameters	
Number of runs	20,000
Number of lorries	100
Number of cities	500
Connectivity	90%
Tax period	5 days
Tax amount	2
Initial wealth	10
Generation wealth	50
Destruction wealth	0
Load generation frequency	1 load every 3 days
Load reward	3
Max loads in city	5

Table 3.11: Load transportation scenario, a stable system: Experiment parameters.

(a) t-test for equality of means

Significance Level	0.05	
	Sample 1	**Sample 2**
Mean	442.1032	442.0002
Variance	130.6010	124.5572
Observations	4001	4001
DoF	8000	
t stat	0.407763224	
$P(T \leq t)$ 2-tail	0.683458432	
t crit. 2-tail	$\pm$1.960261216	
t stat in crit.region	No	
Accept H_0	Yes	

(b) F-test for equality of variance

Significance Level	0.05	
	Sample 1	**Sample 2**
Mean	442.1032	442.0002
Variance	130.6010	124.5572
Observations	4001	4001
DoF	4001	4001
F stat	1.048522604	
F crit. 2-tail lower	0.939893954	
F crit. 2-tail upper	1.063949817	
F stat in crit.region	No	
Accept H_0	Yes	

Table 3.12: Load transportation scenario, a stable system: Statistical analysis on the metric "number of lorries". As the original hypotheses are both accepted, we conclude that the two samples come from the same distribution.

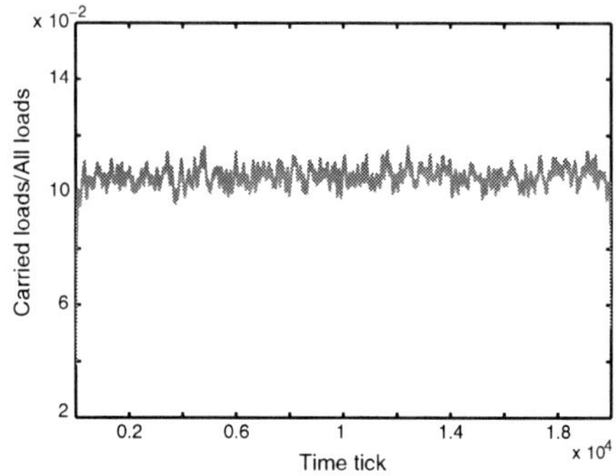

(a) Carried loads ratio

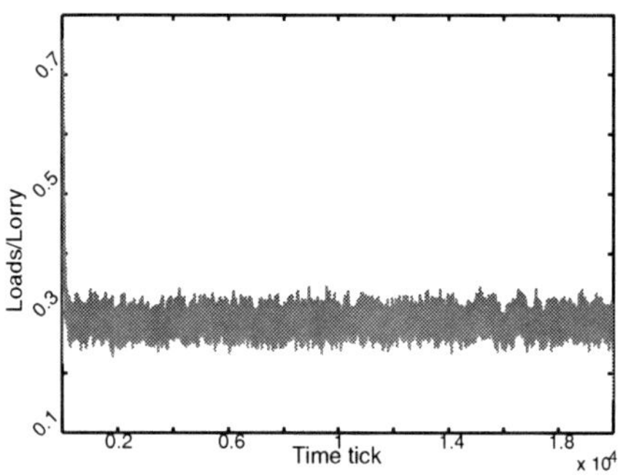

(b) Loads per lorry

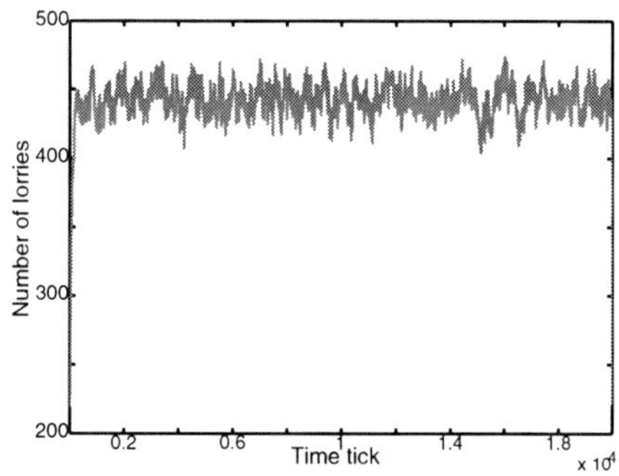

(c) Number of lorries

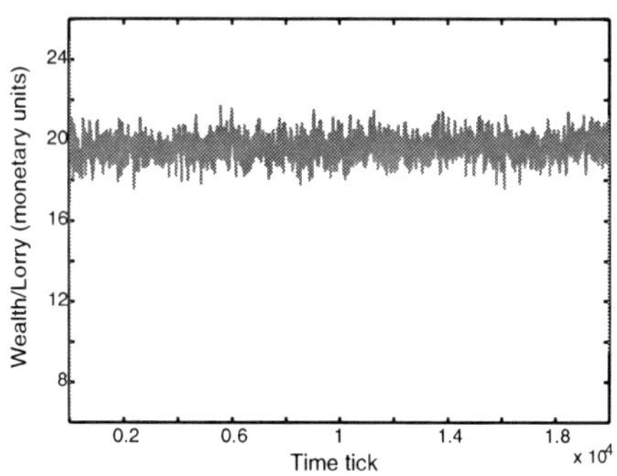

(d) Wealth per lorry

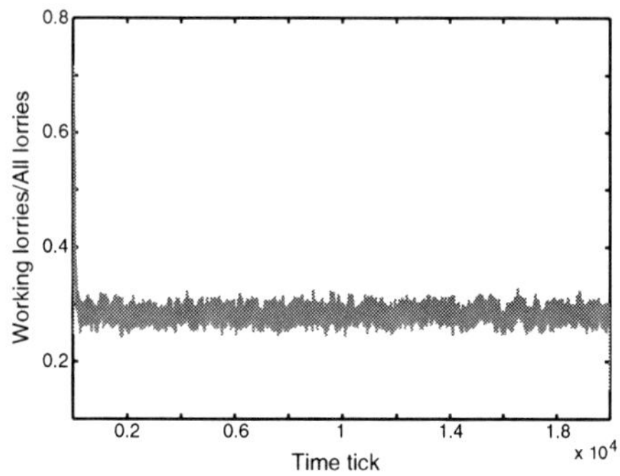

(e) Working lorries ratio

Fig. 3.9: Load transportation scenario, a stable system. All the characteristic metrics reach stationary distributions. The system is therefore considered stable.

The graphs depict a fairly stable system. This observation is verified using statistical hypothesis testing on the characteristic metrics which compared the last 4000 rounds to the 4000 rounds before those. Table 3.12 shows the tests performed on the metric “number of lorries”.

Introducing Perturbations

The initial conditions for this experiment are the same as those for the previous experiment where the system was stable. This time, at time tick 4000, the minimum wealth a lorry has to have in order to be able to generate a new lorry is increased from 50 monetary units to 5×10^7 monetary units, keeping all other parameters the same. The parameters of the experiment are presented in Table 3.13. The graphs associated to the experiment are shown below in Figure 3.10.

Simulation Parameters		**Perturbation Parameters**	
Number of runs	20,000	Start shock at tick	4000
Number of lorries	100	Shock duration (ticks)	1
Number of cities	500	Parameter to shock	Generation wealth
Connectivity	90%		
Tax period	5 days	Shock magnitude (%)	10^8
Tax amount	2		
Initial wealth	10		
Generation wealth	50		
Destruction wealth	0		
Load generation frequency	1 load every 3 days		
Load reward	3		
Max loads in city	5		

Table 3.13: Load transportation scenario, perturbation 1: Simulation and perturbation parameters.

This perturbation will naturally cause the frequency of new lorry generation to decrease. Due to the decrease in the number of lorries, there are more jobs for the remaining lorries to carry out; therefore the working lorries ratio increases dramatically, as is the wealth per lorry ratio and the loads per lorry ratio. However, the remaining lorries do not seem to be enough to cope with the number of loads that are generated in the system. As a result, the carried loads ratio drops after the perturbation occurs. It was shown in the previous experiment that the system is stable under those specific initial conditions. It is evident from the graphs that the load transportation system was driven to instability as a result of the shock that was injected. In this instance, a perturbation caused an originally stable system to become unstable. In another experiment that was performed it was shown that it is possible for the opposite to happen, i.e. a system which is initially unstable can become stable after it has suffered a perturbation.

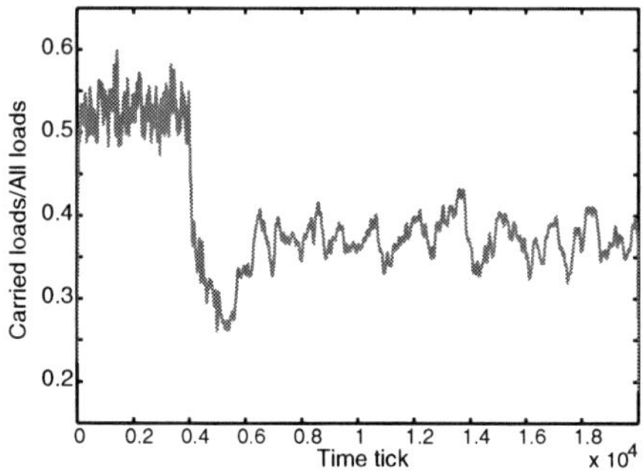

(a) Carried loads ratio

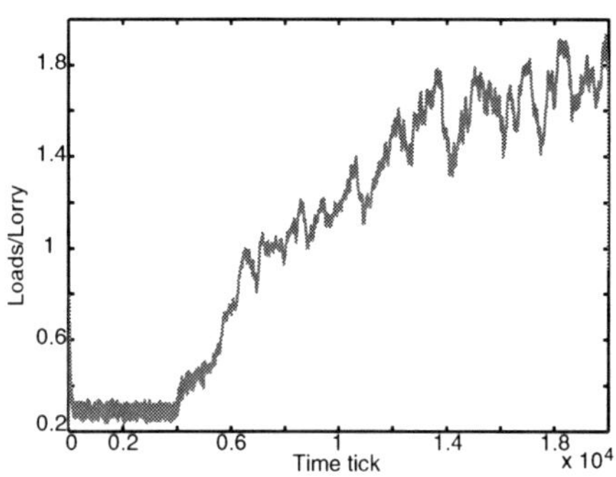

(b) Loads per lorry

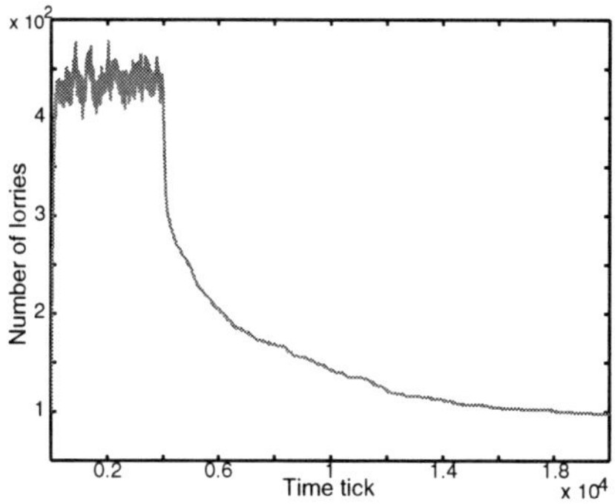

(c) Number of lorries

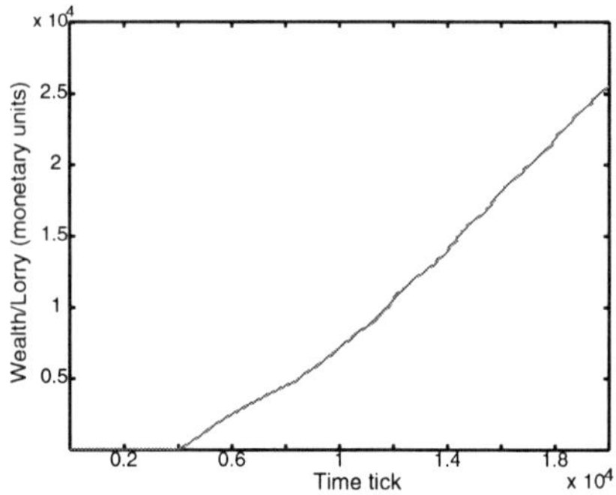

(d) Wealth per lorry

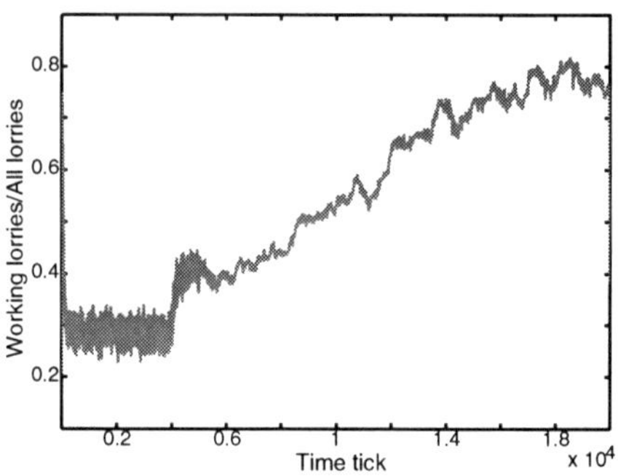

(e) Working lorries ratio

Fig. 3.10: Load transportation scenario, perturbation 1. The perturbation introduced at round 4000 causes the system to destabilise.

The simulation was started with the same initial parameters as the experiment for the unstable system shown above. A shock was injected at time tick 4000. The maximum number of loads in each city was decreased by 80%. The parameters for this experiment are shown in Table 3.14 and the graphs associated to it are shown in Figure 3.11.

As there are now fewer loads in the network of cities, there are fewer jobs for the lorries, hence the decrease in the number of lorries. Consequently, the remaining lorries have a greater workload to cope with. This is apparent in the graph that shows the loads per lorry. This ratio was fluctuating around 50% before the perturbation was introduced and is fluctuating around 57% after. The working lorries ratio increases from 42% to 88%.

Simulation Parameters		**Perturbation Parameters**	
Number of runs	20,000	Start shock at tick	4000
Number of lorries	100	Shock duration (ticks)	1
Number of cities	500	Parameter to shock	Max loads in city
Connectivity	40%		
Tax period	5 days	Shock magnitude (%)	80
Tax amount	2		
Initial wealth	10		
Generation wealth	50		
Destruction wealth	0		
Load generation frequency	1 load every 3 days		
Load reward	3		
Max loads in city	5		

Table 3.14: Load transportation scenario, perturbation 2: Simulation and perturbation parameters.

The load transportation system seems to work much more efficiently after the injection of the shock, as the carried loads ratio is boosted from 15% to 30%. In addition to all these, statistical hypothesis testing confirms that after the system has recovered from the shock, the metrics examined all fall in stationary distributions. Therefore, it is deduced that there is significant evidence that after a small transient phase the system is stable. This experiment demonstrated how an unstable system could be driven to stability after a perturbation has disturbed it.

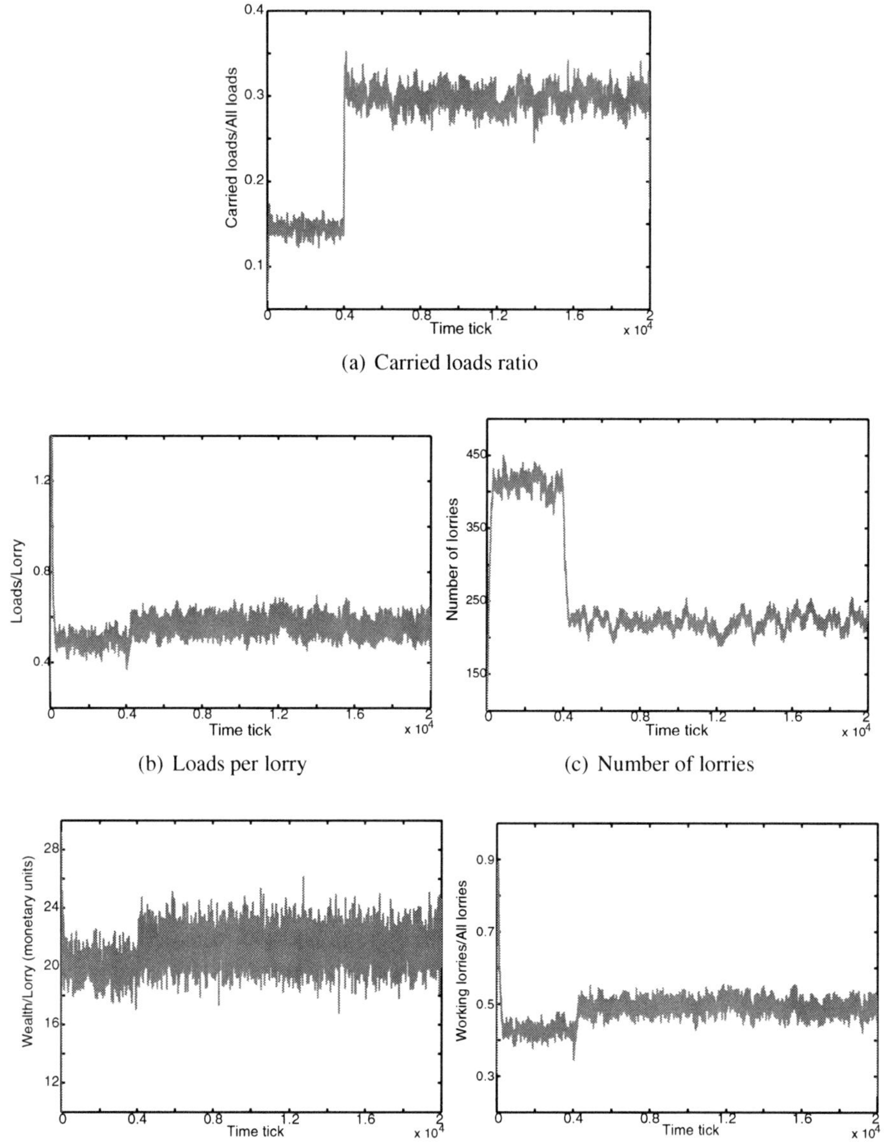

(a) Carried loads ratio

(b) Loads per lorry

(c) Number of lorries

(d) Wealth per lorry

(e) Working lorries ratio

Fig. 3.11: Load transportation scenario, perturbation 2. The perturbation introduced at round 4000, makes the characteristic metrics to fall in equilibrium distributions.

3.4.3 *Virus Spreading Model*

Setting

This model represents a network of nodes arranged in a 2D lattice array. The nodes can suffer from partial or total infection, and can also infect their neighbours. In addition, there is a species of antiviral agents, that traverse this network and can "disinfect" the nodes, at the cost of their own strength.

The whole arrangement is modelled after the immune system of living organisms, but can loosely be applied to a network of interconnected computer hosts that suffer from computer virus infections, as well as mobile agents that roam the network in an effort to stop the spread of the computer viruses. The whole experiment is monitored graphically on a 2D canvas in real time. New infections occur at random points in the network but can also be generated dynamically by the experimenter by clicking on the desired point on the canvas with the mouse. Each node's infection is determined by a "level of infection" parameter that can vary continuously between 0 for an uninfected node and 1 for a fully infected node. Figure 3.12 shows a screenshot of the system with a legend shown in Table 3.15.

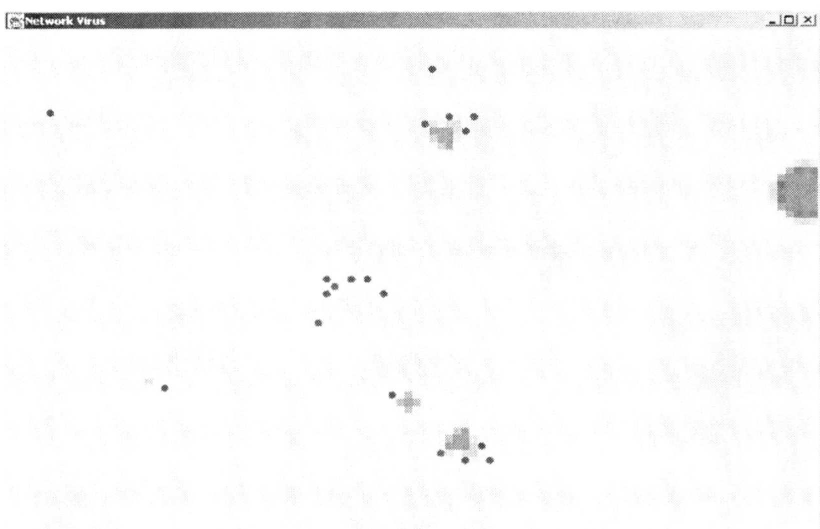

Fig. 3.12: A screenshot from the virus spreading simulation running.

▒ Infected node
• Antivirus walking randomly about the canvas (patrolling mode)
◦ Antivirus walking towards a specific destination. This can interrupt its journey when called by a peer to fight a virus. (travelling mode)
◦ Antivirus on alert (on alert mode)
▒ Canvas (uninfected/healthy node)

Table 3.15: Virus spreading scenario legend.

Interaction

Viruses are generated on random locations of the canvas every certain amount of rounds. Viruses spread to adjacent nodes if not attacked by an antiviral.

An antiviral can be in one of three modes at a time in this scenario. In the first mode it walks randomly about the canvas (patrolling mode), jumping from one location to another. In the second mode, it chooses randomly a specific destination and walks towards that location (travelling mode). Antivirals in either of these two modes will call for help whenever they come across an infection. Only antiviral agents in the second mode that are located within a certain radius (to which we refer as the alert radius) from the point the infection was encountered will be called to help. These will change their mode from "travelling" to "on alert" (the third mode an antiviral can be in). While "on alert" an antiviral travels to the destination where the infection was spotted and cannot be called for help elsewhere. As soon as it reaches the destination, it goes into patrolling mode. When an antiviral has been in patrolling mode for a set amount of time, it then switches to travelling mode.

Antivirals have a *health* attribute and nodes have a *level of infection*,[2] as mentioned above, associated with them. These are the factors that determine who is the winner in a battle between a virus and an antiviral.[3] An antiviral is destroyed when its *health* attribute reaches 0.

The scenario is characterised by the *rate of virus creation* and *the rate of antiviral production*. These are the main elements that affect the model; there are setups for which the infection dominates the canvas and others for which the antivirals manage to control the infections generated. Other important parameters of the model are the

[2] Shown on canvas in a scale from a green to red, green being "no infection" and red being "max level of infection."

[3]
```
int m = Math.min(health, nodeInfectionLevel);
health -= m;
panel.setInfectionLevel(row, col, nodeInfectionLevel-m);
```

rate of virus spreading, the dimensions of the canvas, the initial number of antivirals and the alert radius.

3.4.3.1 Stability Conditions

An obvious situation of stability is that in which the canvas is fully infected (all nodes are red). This is a case of stable equilibrium from which it is not possible to escape. In other words, in this situation the system would be stationary. Another situation that should be considered stable, perhaps more interesting than the previous one, is that in which the total level of infection and the number of antivirals on the canvas do not increase or decrease too much. In fact, according to the proposed definition of stability they should both show evidence that they converge to a distribution after some time the system has been running and remain in this distribution forever.

3.4.3.2 Experiments

Unstable Systems

An experiment was set up on a 70×100 node canvas, initially with 300 antivirals, initially half of them in patrol and half of them in travelling mode. An infected cell contaminates its neighbours (according to the rate of virus spreading) by increasing their level of infection by $\frac{1}{30}$ of its own level of infection at each time tick. New infections start every 100 time steps at random locations on the canvas. New antivirals are created every 400 time steps and the alert radius is 20 nodes. The full list of parameters is shown in Table 3.16 whilst the graphs associated with this experiment are shown in Figure 3.13.

Simulation Parameters	
Number of runs	10,000
Number of antivirals	300
Canvas dimensions	70×100
Rate of virus spreading	$\frac{1}{30}$
Rate of virus creation	$\frac{1}{100}$
Rate of antiviral production	$\frac{1}{400}$
Alert radius	20

Table 3.16: Virus spreading scenario, unstable system 1: Experiment parameters.

The rate of antiviral production is much smaller than that of antiviral creation and as a result the infections cannot be controlled. It is quite clear from the graphs

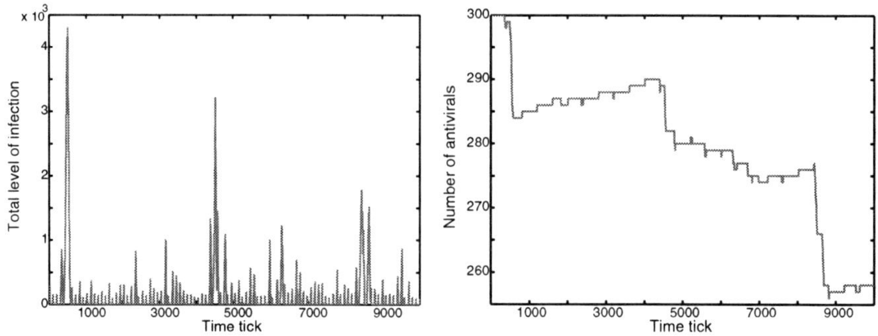

(a) Total level of infection

(b) Number of antivirals

Fig. 3.13: Virus spreading scenario, unstable system 1. The rate of antiviral production is much smaller than that of virus creation, making the infections hard to control and resulting in an unstable system.

of the two metrics that the system is unstable when run under the parameters stated above.

In another experiment that was performed all parameters were kept the same with the exception of the rate of antiviral creation which was made four times bigger than that of virus creation. The parameters of the experiment are set out in Table 3.17 and the relevant graphs in Figure 3.14.

Simulation Parameters	
Number of runs	10,000
Number of antivirals	300
Canvas dimensions	70×100
Rate of virus spreading	$\frac{1}{30}$
Rate of virus creation	$\frac{1}{100}$
Rate of antiviral production	$\frac{1}{25}$
Alert radius	20

Table 3.17: Virus spreading scenario, unstable system 2: Experiment parameters.

It is evident from the graphs that the system is not stable in this experiment either. This observation is verified by the statistical hypothesis tests. The tests performed for the metric "total level of infection" are shown in Table 3.18. The antivirals are still not enough to control the large area of the canvas and bring about stability.

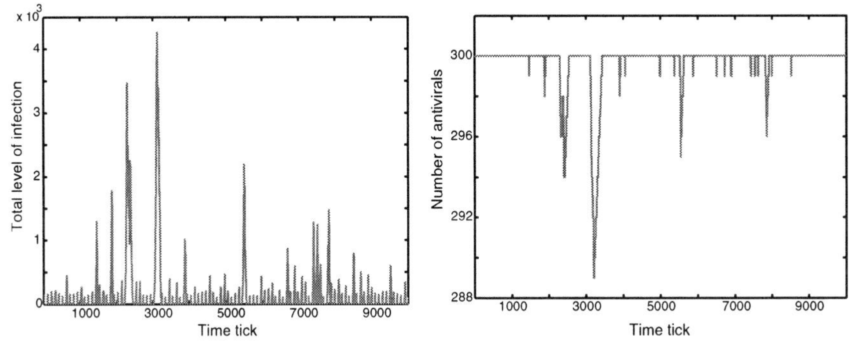

(a) Total level of infection

(b) Number of antivirals

Fig. 3.14: Virus spreading scenario, unstable system 2. Despite the rate of antiviral production being bigger than that of virus creation, the large area of the canvas cannot be effectively controlled.

(a) t-test for equality of means

Significance Level	0.05	
	Sample 1	**Sample 2**
Mean	232.3956	80.60334
Variance	437378.4	38828.61
Observations	4001	4001
DoF	8000	
t stat	13.91349079	
$P(T <= t)$ 2-tail	3.717×10^{-43}	
t crit. 2-tail	± 1.960261216	
t stat in crit.region	Yes	
Accept H_0	No	

(b) F-test for equality of variance

Significance Level	0.05	
	Sample 1	**Sample 2**
Mean	232.3956	80.60334
Variance	437378.4	38828.61
Observations	4001	4001
DoF	4001	4001
F stat	11.26433404	
F crit. 2-tail lower	0.939893954	
F crit. 2-tail upper	1.063949817	
F stat in crit.region	Yes	
Accept H_0	No	

Table 3.18: Virus spreading scenario, unstable system 2: Statistical analysis on the metric "total level of infection." As the original hypotheses are not accepted, we conclude that the two samples do not come from the same distribution.

A Stable System

Another experiment that was performed had significantly different parameters than the ones described above. The canvas was considerably smaller and the initial number of antivirals was much less. Antivirals were produced a hundred times more often than infections. The full parameters of the experiment are shown in Table 3.19 and the related graphs can be found in Figure 3.15.

The statistical tests (Table 3.18) show that the system this time converges to a stationary distribution. It is apparent in the graphs that the number of antivirals remains stationary to what it was set to initially. Even though the infections are generated with a considerably smaller rate than the antivirals, it is evident from the "total level of infection" graph that infection on the canvas was not minor. The fact that the canvas is much smaller than in the previous experiments, makes it more manageable for the antivirals to keep any infections that appear under control.

Simulation Parameters	
Number of runs	10,000
Number of antivirals	5
Canvas dimensions	7×10
Rate of virus spreading	$\frac{1}{50}$
Rate of virus creation	$\frac{1}{100}$
Rate of antiviral production	$\frac{1}{1}$
Alert radius	10

Table 3.19: Virus spreading scenario, a stable system: Experiment parameters.

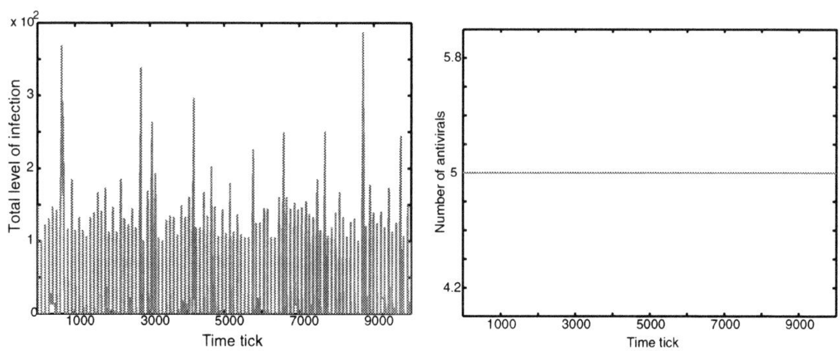

(a) Total level of infection (b) Number of antivirals

Fig. 3.15: Virus spreading scenario, a stable system. The fact that the canvas is much smaller than in the previous experiments, makes it more manageable to the antivirals to keep any infections that appear under control.

(a) t-test for equality of means

Significance Level	0.05	
	Sample 1	**Sample 2**
Mean	19.63059	21.33541
Variance	2450.811	2604.132
Observations	4001	4001
DoF	8000	
t stat	−1.516720579	
$P(T <= t)$ 2-tail	0.129376793	
t crit. 2-tail	±1.960261216	
t stat in crit.region	No	
Accept H_0	Yes	

(b) F-test for equality of variance

Significance Level	0.05	
	Sample 1	**Sample 2**
Mean	19.63059	21.33541
Variance	2450.811	2604.132
Observations	4001	4001
DoF	4001	4001
F stat	0.941123796	
F crit. 2-tail lower	0.939893954	
F crit. 2-tail upper	1.063949817	
F stat in crit.region	No	
Accept H_0	Yes	

Table 3.20: Virus spreading scenario, a stable system: Statistical analysis on the metric "total level of infection." As the original hypotheses are both accepted, we conclude that the two samples come from the same distribution.

Introducing Perturbations

In order to test the response of the system when its normal running is disturbed a series of experiments was performed during which perturbations were introduced in certain parameters. The experiments shown here were started with the same initial parameters as the stable system described above.

For this experiment, at time tick 4000 the infection level of all the cells was set to full (parameter "maximum level of infection"). The graphs in Figure 3.16 show how the system responded to the shock. The parameters of the experiment are in Table 3.21.

Simulation Parameters		**Perturbation Parameters**	
Number of runs	10,000	Start shock at tick	4000
Number of antivirals	5	Shock duration (ticks)	1
Canvas dimensions	7×10	Parameter to shock	Infection level
Rate of virus spreading	$\frac{1}{50}$		
Rate of virus creation	$\frac{1}{100}$	Shock Magnitude (%)	100
Rate of antiviral production	$\frac{1}{1}$		
Alert radius	10		

Table 3.21: Virus spreading scenario, perturbation 1: Experiment parameters.

It is known from the previous experiment that the system is stable when run under those specific initial parameters. It is obvious from the graphs that the perturbation causes a brief period of instability. When a t-test for the means and an F-test for the variances were performed for two samples, Sample 1 and Sample 2 containing data for time ticks 5999 – 7999 and 8000 – 10000, respectively, it was shown that the system did not stabilise after the perturbation. The tests are shown in Table 3.22. This was a case where the introduction of a perturbation causes a stable system to destabilise.

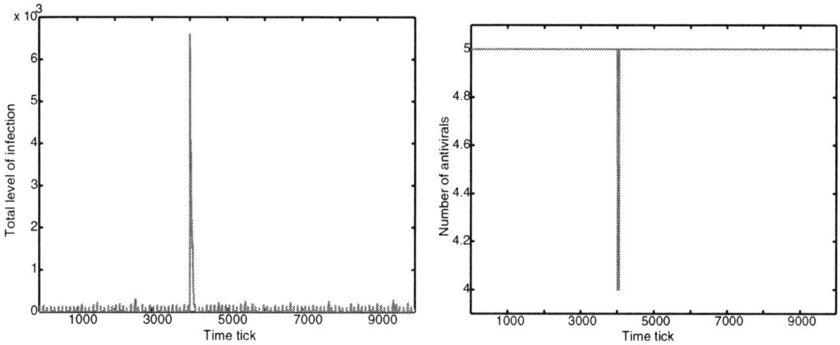

(a) Total level of infection (b) Number of antivirals

Fig. 3.16: Virus spreading scenario, perturbation 1. The perturbation introduced at round 4000 destabilised the system, as is verified by the statistical analysis.

(a) t-test for equality of means

Significance Level	0.05	
	Sample 1	**Sample 2**
Mean	21.68265	22.74162
Variance	2008.611	2566.356
Observations	2001	2001
DoF	4000	
t stat	−0.700347142	
$P(T <= t)$ 2-tail	0.48375188	
t crit. 2-tail	±1.960261216	
t stat in crit.region	No	
Accept H_0	Yes	

(b) F-test for equality of variance

Significance Level	0.05	
	Sample 1	**Sample 2**
Mean	21.68265	22.74162
Variance	2008.611	2566.356
Observations	2001	2001
DoF	2001	2001
F stat	0.782670521	
F crit. 2-tail lower	0.939893954	
F crit. 2-tail upper	1.063949817	
F stat in crit.region	Yes	
Accept H_0	No	

Table 3.22: Virus spreading scenario, perturbation 1: Statistical analysis on the metric "total level of infection." As the original hypotheses are not both accepted, we conclude that the two samples do not come from the same distribution.

In another experiment that was performed the initial parameters were kept the same. However, a different perturbation was introduced. At time tick 4000 all the antiviral agents that were present on the canvas were eliminated. The parameters of the experiment are shown in Table 3.23 and the graphs associated with it in Figure 3.17.

The effect of the perturbation is obvious from the graphs. The antivirals do not get the chance to recover from the shock. The infection spreads throughout the canvas and the level of infection reaches the maximum possible. The number of antivirals oscillates between 0 and 1, and this is not enough to tackle the vast infection that dominates the canvas. Unlike the previous experiment, we see that the system stabilises shortly after the shock.

Simulation Parameters		**Perturbation Parameters**	
Number of runs	10,000	Start shock at tick	4000
Number of antivirals	5	Shock duration (ticks)	1
Canvas dimensions	7×10	Parameter to shock	Number of antivirals
Rate of virus spreading	$\frac{1}{50}$		
Rate of virus creation	$\frac{1}{100}$	Shock magnitude (%)	-100
Rate of antiviral production	$\frac{1}{1}$		
Alert radius	10		

Table 3.23: Virus spreading scenario, perturbation 2: Experiment parameters

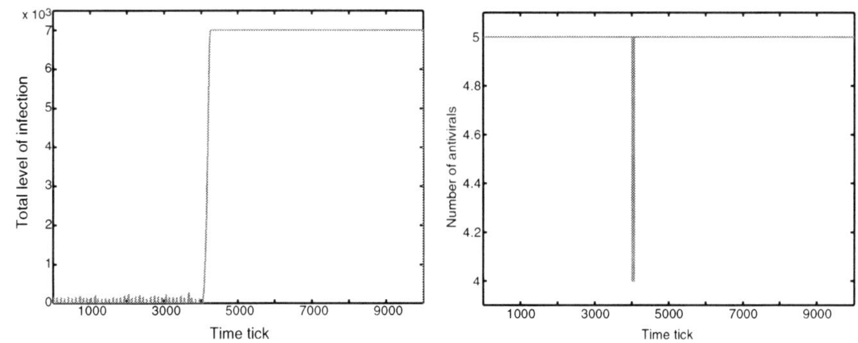

(a) Total Level of Infection (b) Number of Antivirals

Fig. 3.17: Virus spreading scenario, perturbation 2. The system stabilises shortly after the shock introduced in round 4000.

3.4.4 The Market Demonstrator

The market demonstrator was a joint programming effort integrating the research of all the partners of the EEII project consortium. Its aim was to help identify links and interdependencies among the ecology properties openness [Abramov et al., 2003], adaptability [Marques et al., 2003] and stability [Chli and De Wilde, 2003] of multi-agent systems. Each partner programmed the part of the market demonstrator that was related to the the ecology property they were studying. We contributed the code supporting the stability screen, screenshots of which are shown on following pages, as well as some code for the core of the demonstrator.

In this section, the model used for the market demonstrator is described along with the observations from the experiments that were carried out.

3.4.4.1 Setting

The market demonstrator, is essentially a trading environment in which the agents can interact with one another having the role of buyers or sellers in a market. The market system environment is considered to be a set of network hosts where agents can inhabit.

The system uses hybrid agents. Each user agent comprises static and mobile components, referred to as static and mobile agents. The static agent stays on a host and can generate mobile agents to interact with other static or mobile agents. This way, in each host can reside several agents, which can be all buyers, all sellers or mixed.

The static agents have decision support capabilities in order to help users to perform trading tasks or to work in an autonomous way on behalf of a user. The mobile agents can search the world looking for products to buy, or may travel advertising products to sell. These agents are also used to share the knowledge held by the corresponding static agent among other static agents in a voluntary basis.

The mobile agents can access information centers associated to hosts and their residents. The information provided includes:

- identification
- trader's role (buyer or seller)
- products traded

The information collected by the mobile agents is provided to the dispatching static agent to give it a picture of the trading world. Therefore, whenever for example, a buyer static agent wants to buy a specific product it can focus its main search effort on hosts where there is previous information about sellers that trade the product.

Static agents can share some of their knowledge when they talk with mobile agents, by giving and receiving data about the market. The amount and type of the knowledge exchanged is dictated by the level of openness [Abramov et al., 2003] of the system. The knowledge transferred is not necessarily specific about a product.

Mobile agents do not have the same knowledge about the market their corresponding static agents possess, but they are provided with some of that knowledge so they can exchange information with the other static agents. The information collected from the environment by a mobile agent is provided to its static agent on the return of a trip to the market and is integrated with the previously existing data.

Sellers adjust the prices based on supply and demand law principles. They set the price based on the flow of traded products they manage to sell. If products sold are less than a defined reference value the profit margin is reduced. On the contrary, whenever the number of sold products is high then the profit margin is increased. The price adaptation mechanisms are defined by the seller adaptation model [Mariano et al., 2003; Marques et al., 2003].

Any badly performing agent will be eliminated from the system. Sellers will be removed whenever their profit is below 2% and buyers will be removed whenever their performance is below 0.5. After a static agent's removal, there is a time period after which it is replaced, with a certain probability, by a new one. The parameters pertaining to the selling and buying process of the new agent are randomised. The other parameters of the agent do not change. The agent starts with a new wealth amount.

A detailed analysis of the market algorithm can be found in the documentation which accompanies the market demonstrator [Mariano et al., 2003].

3.4.4.2 Stability Conditions

Stability analysis studies how the market system evolves along time. In order to achieve this, a couple of metrics that characterise the system are monitored. These are the tasks executed per trader and the the wealth per trader. Real-time graphs for these metrics are shown in the stability screen of the market demonstrator.

If the market system is stable then the system state will reach a stationary distribution. Metrics such as the tasks per trader and the wealth per trader will reach such a distribution as well. However, the graphs are not enough to check if this has happened. This is the reason statistical hypothesis testing is used.

3.4.4.3 Experiments

In order to explore the parameter space, a large number of experiments was performed for as many combinations of initial parameter values. It is not possible to perform perturbations in the market demonstrator.

For both metrics two samples of size 50 each were taken. One is made up of the latest 50 values of each metric and the other contains the 50 values before those. At each time step we calculate the values of the t-test for the equality of means and the F-test for the equality of variances. If none of them lies in the critical region after a while, then we may say that the metric has converged to a stationary distribution. At the 5% significance level, for two samples of size 50 each, the critical region for

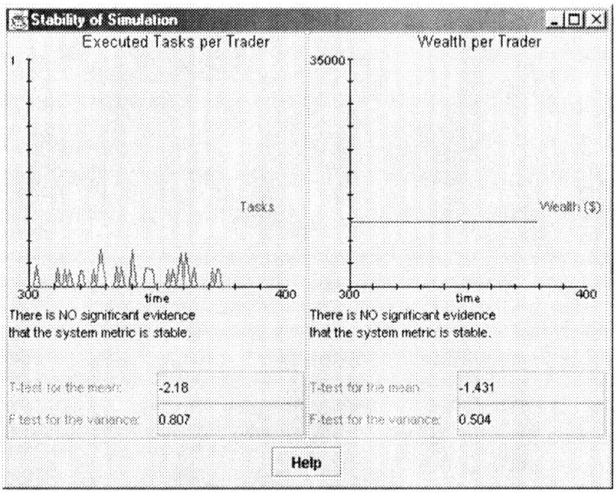

(a) An unstable system. Neither of the metrics has converged to a distribution.

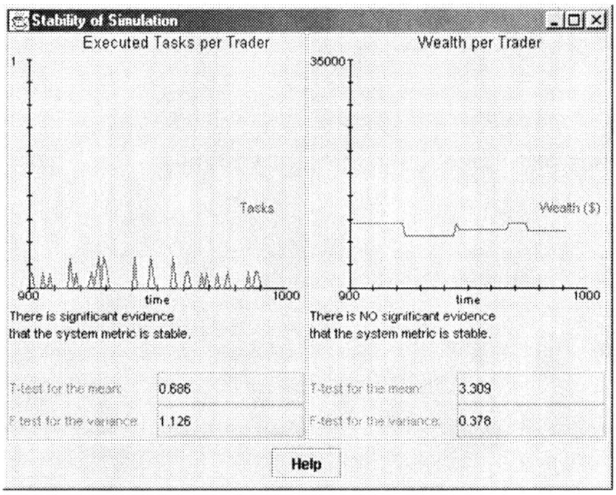

(b) An unstable system. Wealth per Trader has not converged to a distribution.

Fig. 3.18: Market demonstrator. Snapshots of the stability screen depicting unstable systems.

the t-test is $T \leq -1.98$ and $T \geq 1.98$ whereas the critical region for the F-test is $F \leq 0.57$ and $F \geq 1.75$.

When both metrics are shown by the statistical tests to have converged to a stationary distribution then we may say that the system is stable.

Figures 3.18 and 3.19 show a set of snapshots of the stability screen during some experiments performed.

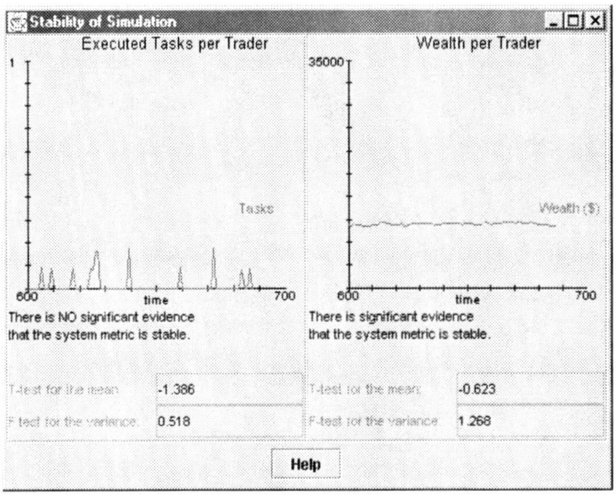

(a) An unstable system. Tasks per Trader has not converged to a distribution.

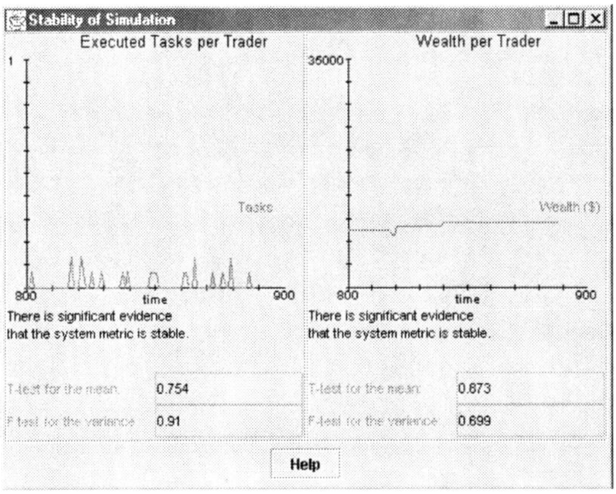

(b) A stable system. Both metrics have converged to a distribution.

Fig. 3.19: Market demonstrator. Snapshots of the stability screen depicting an unstable and a stable system.

3.4.4.4 Discussion

For some of the experiments performed, the system showed evidence of instability, i.e. the two metrics never converged throughout the running of the simulation. For many of the combinations of initial conditions, however, the system metrics eventually converged but only after a long period of instability. In fact, in experiments when the agent adaptability was enabled, periods of convergence were succeeded by periods of instability before the system permanently converged. We conjecture that agent adaptation led to constant change of conditions in the system until a final equilibrium was reached.

Another remarkable interdependency relation was observed when experiments with varying levels of openness [Abramov et al., 2001, 2003] were performed. The higher the openness levels the faster the characteristic metrics converged to equilibrium distributions. Especially, for levels of openness greater than 60% most of the combinations of initial conditions led to stable systems. This is an indication that increasing openness is bound to increase stability.

More work is needed in order to fully investigate the effects that global system properties such as adaptability and openness might have on the stability of a multi-agent system.

3.5 Conclusion

3.5.1 Limitations and Future Work

Our current methodology depends on analysing a multi-agent system through a number of metrics that are representative of its state. The process of finding these metrics at the moment is purely based on our intuition and understanding of the system.

In the future we would like to develop a formal way of discovering the most descriptive metrics of the systems we study.

While this project has provided considerable insight to the stability of ecosystems, it has unveiled a number of paths that deserve further investigation. Hopefully this work will lay the groundwork for further research in the area.

3.5.2 Achievements

This work has been a study of the concept of stability of ecosystems. We began by evaluating definitions of stability that already existed in well-established fields of mathematics. We then explained the reasons these definitions are not suitable for application in the context of multi-agent systems and ecosystems. Subsequently, we proposed a definition of stability which is the only one which is relevant to multi-agent systems with a varying number of agents and is supported by the mathematical framework of stochastic systems.

The study included the design and development of two experimental multi-agent platforms which we analysed, with the intention of illustrating the validity of our definition. Moreover, we introduced statistical hypothesis testing as a means of applying our definition analytically to quantify the stability of complex multi-agent systems.

The study of interdependencies between ecology properties using the market demonstrator unveiled the effect adaptability and openness of the agents have on the stability of a multi-agent system. Agent adaptability led to changing conditions and eventually delayed the convergence of the system. On the other hand, openness, the degree to which agents were willing to exchange information between them, led to faster convergence. This result motivates the work described in the rest of the book. In Chapter 4, agents in a market environment are allowed to adapt and determine themselves whether they will be open to exchange of information and at which degree. It is shown that a straightforward set of assumptions is enough to give rise to exchange. In addition, knowledge exchange is shown to increase the efficiency of the market.

Chapter 4
The Emergence of Knowledge Exchange: An Agent-based Model of a Software Market

Abstract We investigate knowledge exchange among commercial organisations, the rationale behind it and its effects on the market. Knowledge exchange is known to be beneficial for industry, but in order to explain it, authors have used high-level concepts such as network effects, reputation and trust. We attempt to formalise a plausible and elegant explanation for how and why companies adopt information exchange and why it benefits the market as a whole when this happens. This explanation is based on a multi-agent model that simulates a market of software providers. Even though the model does not include any high-level concepts, information exchange naturally emerges during simulations as a successful profitable behaviour. The conclusions reached by this agent-based analysis are twofold: (1) A straightforward set of assumptions is enough to give rise to exchange in a software market. (2) Knowledge exchange is shown to increase the efficiency of the market.

4.1 Introduction

The growth of the Internet as a medium of knowledge exchange has stimulated a lot of scientific interest originating from various disciplines. The willingness of individuals, organisations and commercial firms to share information via the Internet has been remarkable. In some sectors such as scientific research, the communication of newly acquired knowledge and expertise in a field is considered vital for their advancement. On the other hand, in other sectors, the benefits of such exchanges may not be obvious. For instance, it might even be considered damaging for pharmaceutical companies to make public any innovations generated by their research and development (R&D) process. In spite of this view, exchange of intellectual property in some industries occurs quite frequently and in various different ways. These include the forming of strategic partnerships, the participation in open source software projects and the publication of scientific papers by research labs that are part of commercial companies.

M. Chli and P. De Wilde, *Convergence and Knowledge Processing in Multi-Agent Systems*, Advanced Information and Knowledge Processing,

We study the knowledge exchange that occurs in the software industry. In particular, we focus on analysing the rationale behind this exchange as well as its effect on the industry. The complexity of software requirements is a characteristic that distinguishes the software market from others. However, the findings of this work may be relevant to other industries as well. This effort fits within the framework of the digital business ecosystem (DBE) project. The DBE project is an attempt to develop a distributed environment which will interlink European small and medium enterprises (SMEs) that are software providers and foster collaboration between them.

Our broader interest lies in understanding the dynamics of ecosystems (see Chapter 3 and [Chli et al., 2003; De Wilde et al., 2003; Yamasaki and Ushio, 2002]). Furthermore, we are interested in analysing the global system properties which emerge from the interactions that occur in a market ecosystem. We have been using techniques from agent-based modelling to simulate the DBE environment. The main aspects of the DBE market are captured in a model where the SMEs are agents with bounded rationality. This model is then studied using simulations of various settings, and a number of observations are made. One of the most interesting observations is that exchanges between the agents similar to the ones that happen in real-life *arise* in the system. This behaviour *emerges* in the market even though the model does not explicitly account for social issues of trust, network effects or managerial strategies.

The chapter is organised as follows. The following section gives an insight to the digital business ecosystem project and the characteristic of the market that will be developed. In Section 4.3 we sketch the background of this work, namely we review the types of exchanges that occur in markets, giving particular attention to the software market. Section 4.4 details the model used for the investigation carried out. Section 4.5 analyses the experiments performed and the results produced and Section 4.6 concludes.

4.2 Digital Business Ecosystem

In this section we give a brief overview of the digital business ecosystem project, highlighting its aims and motivation. The characteristics of the end-product are identified and special attention is given to the efficiency of the market that will be formed.

4.2.1 A DBE Economy

It is stated in Luck et al. [2003] that virtual organisations make dynamic coalitions of small groups possible. In this way the companies involved can provide more services and make more profits. Moreover, such coalitions can disband when they are no longer effective. At present, coalition formation for virtual organisations is limited, with such organisations largely static.

The overall goal of the DBE project[1] [DBE, 2002] is to launch a new technology paradigm for the creation of a digital business ecosystem that will interlink SMEs and especially software providers. The project is encompassed by the European Union's initiative to become a leader in the field of software application development and to strengthen its SME industry. An open source distributed environment will support the spontaneous evolution, adaptation and composition of software components and services, allowing SMEs that are solution and e-business service providers to cooperate in the production of components and applications adapted to local business needs. This will allow small software providers in Europe to leverage new distribution channels providing niche services in local ecosystems and extending their market reach through the DBE framework. Easy access and large availability of applications, adapted to local SMEs, will foster adoption of technology and local economic growth. It will change the way SMEs and EU software providers use and distribute their products and services.

The main objective of this work, which was carried out as part of the DBE project, was to study the properties of this new type of market. It is clear that the interactions and exchanges between the SMEs within the digital business ecosystem environment will have an effect on the dissemination of information and subsequently to the efficiency of the market.

4.2.2 Market Efficiency

Within the environment of the DBE, business alliances, networks and supply chains require much less effort to be formed. This will promote cooperation and easier dissemination of information between the member SMEs. On the other hand, competition for a share of the market between SMEs will become more direct. It is to be hoped, that these factors will raise the levels of efficiency in the DBE market in comparison to a traditional market. While these aspects of the DBE are very interesting and the subject of future research, this work studies how market efficiency is affected by the exchange of information between SMEs. The experiments carried out on our model, confirm that as the agents engage in more information exchanges between them, with time the market efficiency of the system rises.

Efficient markets theory, as proposed by [Fama, 1970], is a field of economics which seeks to explain the operation of an asset market. Specifically, it states that at any given time, the price of an asset reflects all available *information* [Bodie et al., 2002; Damodaran, 2001]. The efficient market hypothesis implies that it is not generally possible to make above-average returns in the stock market over the long term by trading lawfully, except through luck or by obtaining and trading on inside information.

The DBE environment is different from an asset market, so the definition of efficiency needs to be modified, retaining the spirit of the efficient market hypothesis.

[1] The web page of the project can be found at www.digital-ecosystem.org.

In the model of the DBE used in this work, the market is driven by demand which is fixed and unaffected by the supplied DBE services. In this case the market is efficient if, at any given time, the supply of a service reflects all available information. This means that the services supplied are such that they satisfy the underlying market needs optimally. In other words, the SMEs are not concentrating on catering for some needs while others are left unsatisfied. In an efficient DBE market, *all the needs will be satisfied evenly, assuming that there is equal demand for each of them.* To draw a parallel between the traditional definition of an efficient asset market and the proposed definition for the efficiency of the DBE market, consider the following. In an inefficient asset market, a trading agent can earn excessive returns by buying a particular stock which she believes to be undervalued. Similarly, in an inefficient DBE market a company may make excessive profits by satisfying a need which it knows is not sufficiently satisfied. To invert the argument, in an efficient asset market, asset prices adjust instantaneously and in an unbiased fashion to publicly available new information, so that no excess returns can be earned by trading on that information. Similarly, in an efficient DBE market, the supply of services will adjust immediately to any arising information about the underlying needs.

Cooperation, symbiosis [Jing et al., 1999; Eguchi et al., 2006] as well as the efficiency [Sysi-Aho et al., 2004; Savit et al., 1999] of adaptive multi-agent systems has been studied in the context of the simple games. In Sysi-Aho et al. [2004] no verifiable definition of efficiency is given, whereas in Savit et al. [1999] the system is considered to be in an efficient market phase when all information that can be used by the agents' strategies is traded away, and no agent can accumulate more points than an agent making random guesses would. In the work presented in this chapter, market efficiency, cooperation and competition are studied in the context of a more realistic economic market.

4.3 Background

In this section we list a number of ways in which exchange of knowledge between companies happens in a market, and the rationale for each of them is briefly reviewed. As this work focuses on SMEs that are software providers, we survey the key characteristics of the software industry and the exchanges that take place in this particular market.

4.3.1 Exchange in Economic Markets

In an economic market there are many ways in which the firms engage in exchanges between them. These include the forming of strategic partnerships, the participation in open source software projects and the publication of scientific papers by research companies such as HP Labs and Microsoft Research. In the discussion that follows we will briefly examine the rationale behind these different forms of exchange.

For a strategic partnership to be formed, the partners must mutually benefit from the experience, expertise and talent that all the parties bring to the partnership. There usually is an immediate worthy goal or objective that the partners concerned wish to achieve. For instance, they may wish to operate in a new market, or to bring about a change of leadership in the industry they operate in. Hagedoorn [2002] reports a dramatic rise especially in R&D partnerships, over the past 40 years. These partnerships are mostly limited-time project based collaborations as opposed to long-term alliances. The main motives behind them are reported to be related to cost-cutting as well as risk minimisation whilst the partners attempt to enter new technological areas.

There are a number of reasons behind outsourcing of some operations of a company. Most of these focus on cost management, and revolve around the ability to leverage an outsourcing agent's resources – people, process, infrastructure and intellectual property. Economies of scale can bring significant cost savings in each of these areas especially when the companies concerned are small to medium size (SMEs) [Narula, 2004].

Recent economics and management research has studied the phenomenon of commercial firms contributing to open source projects. The main motive indicated by these analyses is strategic [Grand et al., 2004], as set out in more detail in Section 4.3.2 where the specifics of the software industry are analysed. This seems to be consistent with the fact that it is not the leaders in the industry who engage in open source development, but the followers.

Another form of exchange, which at first might seem counter intuitive, is the publication of scientific papers containing the findings that the research commercial companies perform. It may be argued that it would be in the interest of those companies to keep their innovative work to themselves. Another argument, however, is that by publicising their research they invite others to endorse it, add to it and in effect advance it further. Then, they can use the knowledge acquired by this process to better their products.

The model of a software market that we propose as part of this work is simple in the sense that the agents/firms do not have the ability to reason about complex situations. They cannot make decisions to operate in new markets, or form partnerships in order to change the leadership in the industry. They are not equipped with an understanding of operational costs or of the concept of outsourcing and they cannot devise strategies to undercut their competitors. However, they operate in a capitalistic economy where the best of them succeed whilst the worst perish. They are thus equipped with a simplistic mechanism of reinforcement learning, i.e. being rewarded or punished for choices that prove to be good or bad, respectively. When given the opportunity to engage in exchange of services between them, they learn with time under which circumstances this is beneficial to them and they proceed with it without ever being biased by external factors towards exchanging.

4.3.2 The Software Industry

Complexity is a key characteristic of software which distinguishes the software industry from others. Typical software products carry a large number of features, with innumerable [Bessen, 2006] interactions between them. For a program to be successful in the market, it is necessary that it has the right set of features to satisfy the customer base and that these features operate successfully together.

The market of proprietary software providers/publishers is dominated by large companies, not SMEs. Microsoft Corporation holds the lion's share in the software market with companies such as Oracle, IBM, Hewlett-Packard and Sun follow with smaller shares.[2]

At the same time, the open source[3] movement has been quite successful in developing relatively complex software products such as Linux, Apache or sendmail that are serious competitors of well-established proprietary software [Schmidt and Schnitzer, 2003]. Networks of thousands of volunteers have contributed to these highly complex products. This appears, as it is pointed out in Bessen [2006], to counter the economic intuition that private agents, without property rights, will not invest sufficient effort in the development of public goods because of free-rider externalities.

Lerner and Tirole [2002] justify the volunteers' motivation for contribution to the open source movement as an opportunity to "signal their quality". In other words, the volunteers believe it will enhance their career prospects, as the names of the contributors are always listed in open source projects. Other individual motivations, such as altruism or opportunity to express creativity, are also mentioned.

It is important to point out that in recent years, open source projects have not only received contributions by individuals. There have been organised efforts by firms such as Sun, IBM and others that have endorsed such projects. A survey by Bonaccorsi and Rossi [2004] which was conducted among firms, as well as the account of Gabriel and Goldman [2002] of Sun Microsystems and Grand et al. [2004] list strategic reasons behind the motivation of firms to contribute to open source projects. These reasons include efforts to undercut rival products, gaining a wider tester base for their own products, initiating a gift economy culture between the firm and the open source developer community (where the firm provides the software for free and the community provides debugging or more source code in return) and giving out the software to clients in order to charge for its maintenance and support.

Previous work in this area includes that of Johnson [2002] and Bessen [2006] who have used mathematical models to explain the emergence of the open source

---

[2] The information reflects the year 2002–2003 and was obtained from IBIS World, a strategic business information provider.
http://www.ibisworld.com/snapshot/industry/default.asp?page=industry&industry_id=1239
accessed on 27/05/2005.

[3] In open source software, the source code for a program is made open and available for anyone to screen. There are different open source licenses which prescribe what one is allowed to do with the source code e.g., screen it, interpret it, make changes etc. This is in contrast to proprietary software licenses where the source code is protected by property rights against modification.

initiative. Johnson focuses more on analysing the individual motives and establishing the relationship between the size of the developer base and whether the development goes on. On the other hand, Bessen concentrates on the firm motives for participation in open source initiatives. Bessen models software as a bit string, each bit being a certain feature of the software. In this way the notion that the number of combinations of features grows exponentially with the number of features is captured, depicting the complexity the software can have. He compares open source development with proprietary, prepackaged provision of software and concludes that the two complement each other, recognising that they serve different groups of customers. The latter suits customers with standard, noncomplex software needs, while the former serves customers who have software development capabilities and who need more complex software products.

Bonaccorsi and Rossi [2003] have designed a multi-agent system simulation with which they explore the circumstances for adoption of open source software. They also conclude that proprietary and open source software will coexist in the future. Their model of the diffusion of the two competing streams of software production takes into account issues such as the effect of advertising, network externalities and achievement of critical mass as in Loch and Huberman [1999].

The stylised model presented in this work simulates a market in which the companies try to satisfy a set of underlying software needs with the services that they develop. The companies follow simple, high-level rules imposed by a capitalistic economy. Interestingly, exchanges between the agents similar to the ones that happen in real software markets, arise in the system. This behaviour *emerges* in the system even though we have avoided modelling issues such as social or strategic motives of the contributors or network effects.

4.4 An Agent-based Model of the DBE

4.4.1 Agent-based Modelling

A new generation of computing techniques analysed in Goonatilake and Treleaven [1995], commonly known as "intelligent systems", has been applied to a variety of financial and business modelling tasks. These techniques include genetic algorithms, neural networks, dynamical systems models, expert systems, rule induction, fuzzy systems and hybrids of these techniques.

Agent-based modelling has been recently used in economics research work to study models of markets, e.g. the Santa Fe artificial stock market [LeBaron, 2002; Bonabeau et al., 1999], and their characteristics [Kirman and Vriend, 2001], in computing-economics interdisciplinary work to study information economies of autonomous agents [Kephart et al., 1997; Kephart, 2002; Griffiths and Luck, 2003; De Wilde, 2004; Sim and Wong, 2001] and business processes [Huang, 2001], in social sciences to study emergent behaviour [Epstein, 2002], issues of trust [Falcone and Castelfranchi, 2001] and to perform syndromic behaviour surveillance [Carley et al., 2006] and in other disciplines.

Much research in multi-agent systems explores how refinements to one agent's reasoning can affect the performance of the system [Brenner, 2004]. Significant effort has been directed towards formally defining emergence in agent-based systems. According to Bar-Yam [2004] a strong emergent property is a property of the system that cannot be found in the properties of the system's parts or in the interactions between the parts. Additionally, in Van Dyke Parunak et al. [2004] the notion of universality is studied: systems whose elements differ widely may have common emergent features.

Agent-based modelling according to Tesfatsion [2005] "is a method for studying systems exhibiting the following two properties:

1. the system is composed of interacting agents; and
2. the system exhibits *emergent* properties, that is, properties arising from the interactions of the agents that cannot be deduced simply by aggregating the properties of the agents."

In models such as the one proposed below, where the interaction of the agents is determined by past experience and the agents continually adapt to that experience, mathematical analysis is typically very limited in its ability to derive the dynamic consequences. In this case, agent-based modelling may be the only practical method of analysis.

We follow a "bottom-up" approach, after a brief overview of the methods used in Section C which follows, in Sections 4.4.2 and 4.4.3 we describe the first principles of agent behaviour and in Section 4.5 we analyse the macro-properties emerging from the agent interactions.

4.4.2 The Setting

In this section, the model used for the simulation of the DBE environment is set out. SMEs are modelled as agents in a multi-agent system. The services the SMEs provide are modelled as bit strings in the same manner software services are modelled in Bessen [2006], each bit symbolising a feature of the service. Finally, the underlying market is modelled by a set of requests (market needs) which are exogenous and are generated randomly. A request is a bit string of the same size as a service bit string.

Each SME has a population (or portfolio) of services. This population is not static throughout the lifetime of the SME. If a service is successful, the SME tends to add similar services to the portfolio while an unsuccessful service is usually discarded. The whole process is modelled quite elegantly by a genetic algorithm (GA) within the portfolio which involves mutation and crossover with survival of the fittest. Through this population each SME can choose which request it will try to satisfy. The genetic algorithm represents the R&D businesses perform in order to improve their services. An overview of genetic algorithms is given in Appendix C.

The use of genetic algorithms is a natural and simple way to model R&D, with minimal assumptions. However, other algorithms can be used in place of the GA to model R&D. The GA captures the following characteristics:

1. trying to find a solution to a particular problem,
2. using a population of possible solutions.

Any other method that can capture the above two characteristics may be used in place of the GA.

The objective of an SME is to increase its fitness. Each SME maintains a portfolio of candidate services, only one of which will be submitted to the market. Each candidate service receives a rating according to how profitable it would be for the SME if it was submitted to the market. This calculation is performed using the services submitted by all other SMEs in the previous round. The rating of each candidate service within the SME portfolio is used to: (a) decide on which service to submit to the market and (b) evolve the best services in the portfolio (with mutation and crossover) and eliminate the worst services.

The fitness of a service measures how profitable it is to its owner. The profitability of a service depends on:

1. how close the service is to the market needs (service-request similarity) and
2. how many other services satisfy those needs (limited demand) .

The fitness of an SME equals the fitness of the service it offers.

In the section that follows we discuss the factors that affect the fitness (or profitability) of a service.

4.4.2.1 Service-Request Similarity and Limited Demand

Assume there are m SMEs in the market, each one offering a single service. Consider a service S and a request R, each represented by a bit string of fixed length. Similarity is measured by the percentage of shared bit values between S and R, denoted by $d(R_i,S_j), 0 \leq d \leq 1$. If the market requests are $R_1, R_2, ..., R_n$, services in the market are $S_1(t), \ldots, S_m(t)$, the fitness of a service $S_j(t)$ is

$$U_j(t) = \sum_{i=0}^{n} (\phi(R_i, S_j(t)) \times \rho_i(t)), \tag{4.1}$$

where

$$\phi(R_i, S_j(t)) = e^{-\frac{1-d(Ri, S_j(t))}{\alpha^2}}. \tag{4.2}$$

The variable ϕ is used to parametrise the fitness landscape (make maxima more or less pronounced), α being a shape parameter. Figure 4.1 shows the relationship of ϕ with with the similarity d. The weight/discounting factor ρ is given by

$$\rho_i(t) = min\left\{1, \frac{1}{\sum_{j=1}\phi[R_i, S_j(t)]}\right\}. \tag{4.3}$$

The variable ρ models the fact that the demand in the market is limited. When a request is saturated (i.e. too many services try to satisfy it) then $\rho < 1$. Subsequently, the fitness of the service is discounted. Otherwise, when $\rho = 1$ the fitness of the service equals ϕ.

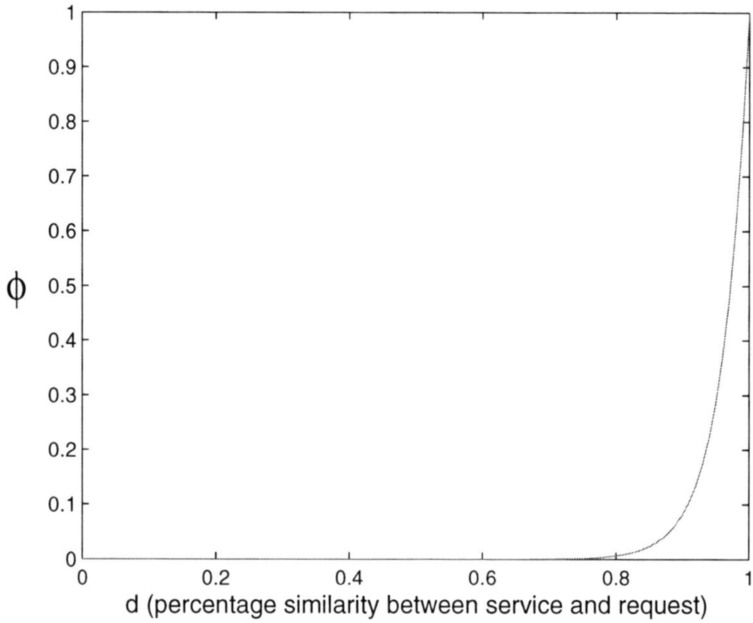

Fig. 4.1: The relationship of ϕ with the service-request similarity d for $a = 0.2$. The variable ϕ is used to parametrise the fitness landscape (make maxima more or less pronounced).

The fitness of an SME is equal to the fitness of the service it submits to the market.

4.4.2.2 Satisfaction of Requests and Market Efficiency

An additional useful measure is the degree to which a request is satisfied. This is a metric of how saturated it is , in terms of how many services try to satisfy it and how similar their features are to those of the request. The degree of satisfaction $Q_i(t)$ of a request R_i at round t is given by:

$$Q_i(t) = \sum_{j=1}^{m} \phi[R_i, S_j(t)]. \tag{4.4}$$

This measure is necessary for assessing the efficiency of the DBE market. As discussed in Section 4.2.2, in an efficient DBE market all the market requests will be equally saturated , assuming there is the same demand for all of them. Thus, we calculate the standard deviation $\sigma(t)$ of the satisfaction values of all the requests in the market at round t. The smaller it is, the more similar to each other the saturation levels of the requests are.

$$\sigma(t) = stdev\{Q_1(t), \ldots, Q_n(t)\} \tag{4.5}$$

The mean of the saturation values will be constant due to the demand in the model being fixed.

4.4.3 Exchange of Services

As outlined in 4.3.1 exchange of services may encompass many real-life situations that occur in a market. These include the forming of strategic partnerships of companies, outsourcing of operations, participation in free/open source projects and others. The setting described here is a loose model of such situations which aims to identify the basic factors that lead to this general behaviour of exchanging. In our model, the exchange involves selecting a set of services from one SME's portfolio and swapping them with the corresponding set of services of the other SME's portfolio. When a company chooses to swap a set of services, this means that after the exchange has taken place it won't have these services in its portfolio any more. The services in a portfolio of a company are sorted according to their fitness (i.e. how profitable they are to the SME that owns them). The model in its current state supports exchange of services that are in the same rank, in the two portfolios, e.g. the 5th service in the portfolio of one SME with the 5th service in the portfolio of the other.[4]

At each time tick, the SMEs need to decide whether they want to exchange some of their services with one of the other SMEs. A statistical classification algorithm is used to model the decision problems an individual agent faces. An overview of statistical classification is given in Appendix C.

4.4.3.1 Exchange Decisions

Every SME has a classifier system which it uses to decide on whether they want to exchange some of their services with one of the other SMEs. The rules of the

[4] Experiments have shown that the rank of the services being exchanged is not of much significance, assuming that services of the same rank are being exchanged, but we plan to investigate this further in the future.

classifier are shown in Table 4.1 below. The objective of an SME at all times is to increase its fitness. The rules' condition part refers to the rank of the SME in the market with respect to the rank of its colleagues. The action part examines the potential partner's rank and prompts the SMEs either to engage in an exchange with a specific type of partner or abstain from exchanging. For simplicity, the SMEs are clustered in three[5] groups according to their rank. Therefore we have upper, middle and lower-ranked SMEs. For an exchange to take place both parties need to agree.

We experiment both with settings in which the rank is based on the fitness of each company and others where the rank is not linked to SME performance in any way. For example, in experiments where rank is based on SME performance, the SME with the highest fitness will have $rank = 1$, whilst the SME with the lowest fitness will have $rank = number\ of\ SMEs$. On the other hand, in experiments where rank is unrelated to performance in the market the rank of an SME may be its id number. In Section 4.5 we analyse these experiments and present the effect the different meanings rank may take have on the learning that occurs.

`if` *my rank* = *lower* `then`	*exchange with lower cluster,*	s_1
`if` *my rank* = *lower* `then`	*exchange with middle cluster,*	s_2
`if` *my rank* = *lower* `then`	*exchange with upper cluster,*	s_3
`if` *my rank* = *lower* `then`	*do not exchange,*	s_4
`if` *my rank* = *middle* `then`	*exchange with lower cluster,*	s_5
`if` *my rank* = *middle* `then`	*exchange with middle cluster,*	s_6
`if` *my rank* = *middle* `then`	*exchange with upper cluster,*	s_7
`if` *my rank* = *middle* `then`	*do not exchange,*	s_8
`if` *my rank* = *upper* `then`	*exchange with lower cluster,*	s_9
⋮	⋮	⋮

Table 4.1: A few example rules of the classifier which an SME uses to decide on what type of partner to choose for an exchange.

The classifier system operates as follows [Kirman and Vriend, 2001]. First, it examines the `if` part of each rule to determine and shortlist the rules whose conditions are satisfied at a given time t. It then assigns a score b to the shortlisted rules, s_k being the strength of the kth rule:

$$b_k(t) = s_k(t) + \varepsilon, \text{where } \varepsilon \simeq N(0, \sigma). \tag{4.6}$$

The rule with the highest score b becomes the *active rule*.

After the active rule has been executed and has generated payoff ω during the previous round $t-1$, the classifier system updates its strength s:

[5] Experiments have been carried out which showed that model behaviour doesn't vary significantly with cluster size. Three is the optimal number of clusters with respect to having a model which is realistic enough while taking a reasonable amount of time to execute and giving us the ability to present the results in an efficient and clear way.

$$s_k = s_k(t-1) - cs_k(t-1) + c\omega(t-1), \text{where } c \in [0,1]. \quad (4.7)$$

In other words, $\Delta s_k(t) = c[\omega(t-1) - s_k(t-1)]$. Therefore, as long as the payoff in round $t-1$ is greater than the strength of the rule on that round, the strength will increase. If the selection of the rule led to a small payoff being generated, the strength of the rule will decrease, making it less likely to be activated in the future. The strength of each rule converges to some weighted average of the rewards ω generated by the environment in response to that specific rule.

In our implementation of the model all the rules have initial strength 0. The rule strengths are adjusted as the simulation goes on. The strength of each rule that is activated is updated at every round using the following payoff from the external environment: $\omega(t) = U_j(t) - U_j(t-1)$. In other words, the payoff is the difference in the fitness of the company between the current and the previous round. The payoff may be negative, zero, or positive according to the change in fitness.

4.4.3.2 Exchange Decisions Resolution

Once the companies that have decided to participate in an exchange have selected the type of partner they prefer, they are teamed up accordingly. For instance, an SME in the cluster of middle-ranked SMEs, who has decided to exchange with a high fitness company will be coupled with a high-ranked company who wants to exchange with a middle-ranked one. If a suitable partner is not found the exchange does not happen. The strength of the rule that was activated in that case will still be updated even if the transaction was not carried out. This reflects the effect choosing a partner who is unwilling to collaborate has on the fitness of the company.

4.4.4 Discussion

The model outlined above is simple in that it has captured the main aspects of a digital business ecosystem. It is the model of a market in which the companies try to satisfy a set of underlying requests. They do so by producing and making available services that are as close as possible to the specified requests. Each company has its own R&D portfolio of services that it evolves. At each round the companies go to the market with what they believe is the best service in their portfolio. In addition, the companies have an option to exchange services with partners that they select themselves.

The simplicity of the model is also inherent in the behaviour of the agents. The agents have to find which is the best service to make available, based on the services that were submitted to the market during the previous round. Also, they need to decide whether and with whom to exchange their services based on their rank in the market. These are all abstractions from reality. We do not assume any network effects in the market. Also, there are no indicators about value of the brand of a company.

4.5 Analysis of the Model

In this section the experiments carried out using the model of the DBE are described. The analysis focuses on two main findings:

1. The companies discover themselves that under certain circumstances it is beneficial to them to exchange services between them.
2. Allowing exchange to take place in the market, makes for greater market efficiency levels.

It is important at this point to stress that the choice to exchange services is not a practice that is imposed by the model mechanism. Instead, it is a feature that emerges from the classifiers as it is a gainful practice for the companies under certain circumstances.

The model behaviour is quite general and has been observed for a very wide range of parameters and initial conditions. The graphs and figures shown below come from randomly selected runs of the simulation, unless it is stated otherwise.

4.5.1 Service Exchange

4.5.1.1 Exchange Decision

As described in Section 4.4.3 each agent/company uses a classifier to decide whether or not to exchange some of its services. The decision is based on the company's rank in the market. Figures 4.2(a) and 4.2(b) show the average strength of the rules of all the companies' classifiers at the end of a simulation which lasted for 10,000 iterations. The companies are ranked according to their fitness. The fittest company will have rank 1 whilst the least fit company will have rank equal to the number of companies in the market. To make for less time-consuming simulations and more readable graphs the companies are grouped into three clusters according to their rank; so they are divided into lower-, mid- and upper- ranked SMEs. Figure 4.2(a) was generated from a run of the simulation where the DBE market consisted of 21 SMEs, each having 20 services in its portfolio. Each service had ten features. There were four software requests in the market, generated randomly. The run of the simulation which produced Figure 4.2(b) had largely similar parameters, the difference being that there were 30 services in the SMEs' portfolios and there were five requests in the market.

The strongest of the rules at each situation is the one which is more likely to be activated. In other words, it is shown in Figures 4.2(a) and 4.2(b) that if a company belongs to the mid or lower cluster it is likely that it will choose to participate in an exchange (preferably with a upper-ranked company) while if it belongs to the upper-ranked cluster it will avoid engaging in any exchange activities. The graphs show that in the less successful, lower-ranked SMEs the classifier rules that correspond to `exchange` actions have higher strengths than the rule that leads SMEs not to exchange. The opposite holds for higher-ranked SMEs, i.e. the rule that corresponds to a `not exchange` action has higher strength than the `exchange` rules. For mid-ranked SMEs, a rule prompting the firm to exchange is the stronger of all, but exchanging is not always a profitable practice; the rule that leads the SME to avoid exchanging is often stronger than some `exchange` rules.

The generality in the behaviour of the model is confirmed by Figure 4.2(c). A wide range of parameters and initial conditions were varied in a total of 200 experiments, keeping the number of SMEs in the market constant (21). Figure 4.2(c) shows the average values of the SME classifiers' strengths over those 200 experiments. The general trend which emerges is that the average performing (mid cluster) and worst performing (lower cluster) SMEs learn that it is to their advantage to exchange services with others while the top performers (upper cluster) learn to avoid exchanging .

To understand better the behaviour of the system we performed experiments with different rankings of the SMEs. Amongst the ranking methods we tested were variants of the fitness ranking, as well as rankings unrelated to SME performance altogether. The results seem to indicate that information exchange emerges as long as the ranking is in some way related to SME performance. We show in Figure 4.3(a) the rule strengths in the case the SMEs were ranked according to fitness growth rates

$$\Delta U_j(t) = U_j(t) - U_j(t-1), \tag{4.8}$$

rather than fitness itself. The graphs produced are similar in pattern to those in Figure 4.2(c). These strengths imply that the rules are significant and learning has taken place in the system. Similar results, shown in Figure 4.3(b), were produced when SMEs were ranked according to the N-moving average of their fitness, given by

$$\mu = \frac{1}{N} \sum_{T=t-N}^{t} U_j(T). \tag{4.9}$$

On the other hand, in Figure 4.3(c) a typical case of a ranking that is unrelated to SME fitness is shown. In that particular case we gave the SMEs an arbitrary ranking that remained fixed throughout the simulation. The rule strengths indicate that no rule is significantly more important than any other one implying that the rules are not relevant and no learning has occurred. We also tried a completely random and constantly changing SME ranking which produced similar results, shown in Figure 4.3(d).

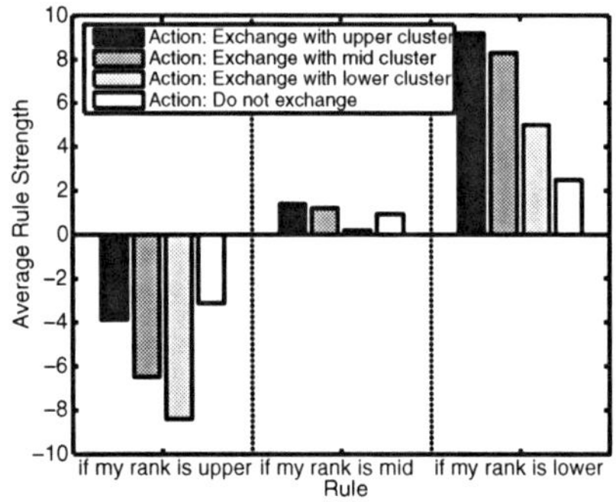

(a) Run 1

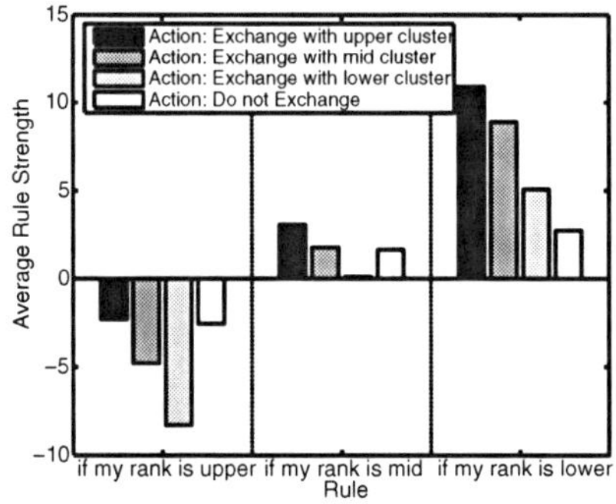

(b) Run 2

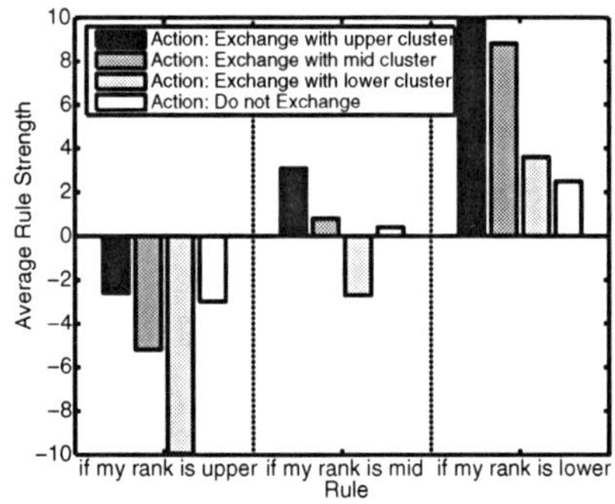

(c) Average over 200 experiments

Fig. 4.2: Average exchange rule strength – 1. The graphs show the strength values of each rule at the end of a simulation averaged out over all SMEs' classifiers. The SMEs decide whether to participate in an exchange of services according to their rank. The classifier each SME has is as follows:

`if` *my rank* = *lower* `then` *exchange with lower cluster*, s_1
`if` *my rank* = *lower* `then` *exchange with middle cluster*, s_2
`if` *my rank* = *lower* `then` *exchange with upper cluster*, s_3
`if` *my rank* = *lower* `then` *do not exchange*, s_4
`if` *my rank* = *middle* `then` *exchange with lower cluster*, s_5
`if` *my rank* = *middle* `then` *exchange with middle cluster*, s_6
`if` *my rank* = *middle* `then` *exchange with upper cluster*, s_7
`if` *my rank* = *middle* `then` *do not exchange*, s_8
`if` *my rank* = *upper* `then` *exchange with lower cluster*, s_9

⋮ ⋮ ⋮

For Figures 4.2(a) – 4.2(c) and 4.3(a) – 4.3(b) the rank of the SMEs is based on measures related to their fitness, while Figures 4.3(c) and 4.3(d) were created for settings in which the SME rank was unrelated to fitness. The graphs show that in settings where the rank is associated with some fitness measure the SMEs that are further down in the rank learn that is beneficial to them to participate in an exchange.

4.2(a) Run 1 parameters: 21 SMEs, each having 20 services in its portfolio. Each service had ten features. There were four different software requests in the market.

4.2(b) Run 2 parameters: 21 SMEs, each having 30 services in its portfolio. Each service had ten features. There were five different software requests in the market.

4.2(c) Average values over 200 experiments This figure confirms the generality of the behaviour of the model. A wide range of parameters and initial conditions were varied in a total of 200 experiments, keeping the number of SMEs in the market constant (21). The rank was based on the fitness value of the SME.

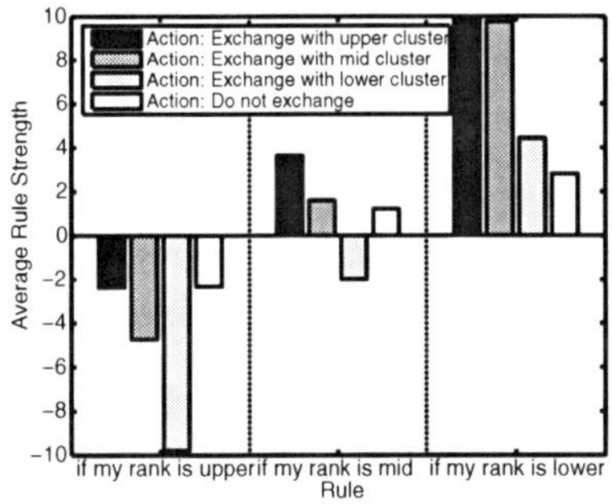

(a) Rank based on fitness growth rate

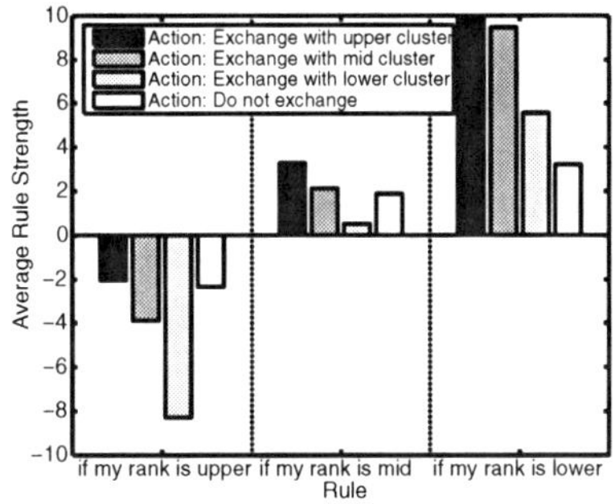

(b) Rank based on fitness moving average

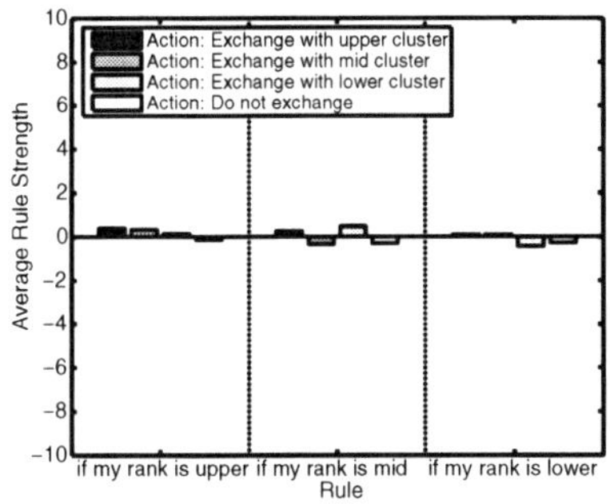

(c) Rank based on SME id which is random and static throughout the simulation

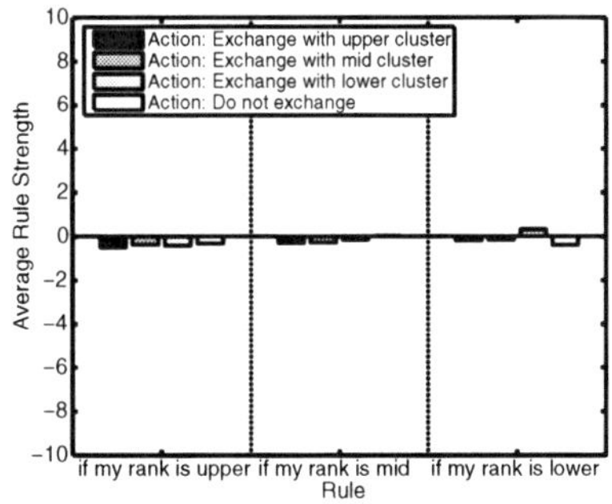

(d) Rank based on SME id which is random and constantly changing throughout the simulation

Fig. 4.3: Average exchange rule strength – 2. The graphs show the strength values of each rule at the end of a simulation averaged out over all SMEs' classifiers. Each SME decides whether to exchange services with another according to its rank. The classifier each SME uses is explained in the caption of Fig. 4.2 and in Section 4.4.3 For Figures 4.2(a) – 4.2(c) and 4.3(a) – 4.3(b) the rank of the SMEs is based on measures related to their fitness, while Figures 4.3(c) and 4.3(d) were created for settings in which the SME rank was unrelated to fitness. The graphs show that in settings where the rank is associated with some fitness measure the SMEs that are further down in the rank learn that is beneficial to them to participate in an exchange. **4.3(a) and 4.3(b) Average exchange rule strength based on SME performance measures.** The SMEs decide on whether to participate in an exchange of services according to their performance (fitness). In Figure 4.3(a) the performance measure deciding the rank of the SMEs is their fitness growth rate, while in Figure 4.3(b) it is the 20-moving average of the SME fitness. As long as the ranking of the SMEs is performance related, information exchange always emerges as a gainful strategy. **4.3(c) and 4.3(d) Average exchange rule strength not based on SME performance measures.** In Figure 4.3(c) the SMEs decide whether to participate in an exchange of services according to their unique id. In Figure 4.3(d) the ranking of the SMEs is random and constantly changes. In both cases, the ranking is unrelated to SME fitness or any other performance measure. The rule strengths indicate that no rule is significantly more important than any other one implying that the rules are not relevant and no learning has occurred.

4.5.1.2 Choice of Exchange Partner

An interesting result which arose from the experiments is the choice of potential partners for the companies who decide to exchange. In all three situations (`if my rank is upper`, `if my rank is mid` and `if my rank is lower`) the strength of the rules that prompt SMEs to exchange reveal a decreasing preference from left to right between upper-, mid- and lower-ranked partners. That result is entirely intuitive and confirms the validity of the model.

A result that may not be so obvious is the fact that the lower-ranked SMEs benefit from exchanging even between themselves. This is reflected in the fairly high strength of the relevant rule and it is better illustrated in Figure 4.4.

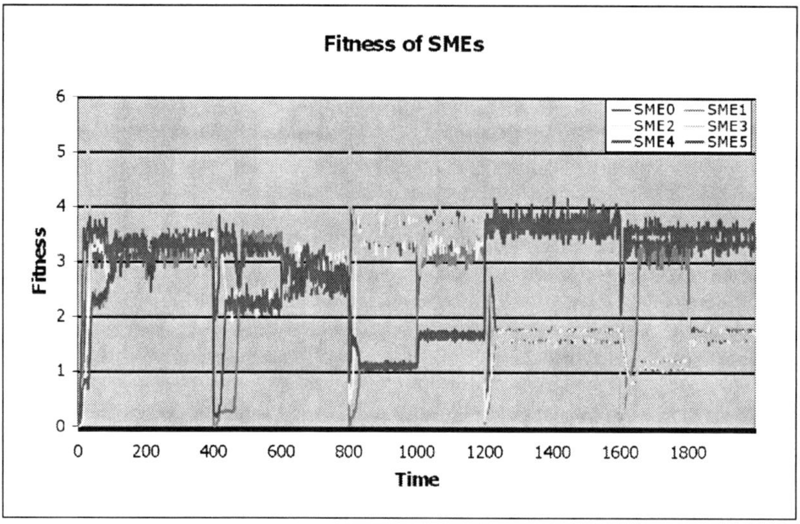

Fig. 4.4: This is an experiment that illustrates that exchange among lower-ranked SMEs is beneficial to them. Every 400 rounds the underlying requests in the market change. Every 200 rounds (but not when the requests change), the lower-ranked SMEs exchanged services between them. In most instances the exchange drives the under-performers up, in terms of fitness.

The experiment that yielded Figure 4.4 is as follows. To make for a more intelligible graph, there are only six SMEs in the market and two distinct requests. Every 400 rounds the underlying requests in the market change. Every 200 rounds (but not when the requests change), the lower-ranked SMEs exchanged services between them. As the purpose of this experiment was to verify the finding that exchange among lower-ranked SMEs is beneficial, the exchange was done deliberately and not using the classifier. As shown in Figure 4.4, in round 200 the exchange does not upset the equilibrium too much as the SMEs have more or less the same fitness. In round 600 the exchange drives the lower-ranked SMEs up, whilst damaging the fitness of the others in the market. In round 1000 the exchange not only drives the under-performers up but also causes one of them, SME_1 to join the upper cluster.

The experiment described above illustrated that exchanges between low-ranked SMEs can be highly beneficial. This is because the fusion of their portfolios may yield services that enable them to operate in a new market segment, in other words it may lead them to satisfy another request which was previously not catered for. This can cause their rank in the market to improve and even bring about a change of leadership in the industry.

4.5.2 Market Efficiency

As discussed in Section 4.2.2, the increased flow of information within the DBE, will make it easier for the participating companies to find the right trading partners. Consequently, it will make for greater market efficiency levels in comparison to a conventional market (e.g. the software industry). An interesting observation which emerged from the analysis of the simulations carried out is that allowing the SMEs to exchange services between them, increases the efficiency further.

A DBE market is considered efficient when all the requests are equally saturated. In an efficient DBE market, the supply of services will adjust immediately to any arising information about the underlying requests. In other words, there is no excess profit to be gained by an SME choosing to satisfy another request than the ones it currently does. As mentioned in Section 4.4.2.2, the degree of satisfaction of a request R is given by Equation (4.4). In order to assess the level of efficiency in the market we need to calculate the standard deviation $\sigma(t)$ of the satisfaction values of all the requests in the market, as given by Equation (4.5). The smaller it is, the more similar to each other the saturation levels of the requests are. It is important to mention at this point that the mean of the saturation levels remains constant, because in the model we assume equal demand for all of them, and it is equal to $\frac{\text{number of services in the DBE}}{\text{number of requests}}$.

Figure 4.5 shows the standard deviation $\sigma(t)$ of the saturation values $Q_i(t)$ of all the requests $\{R_1, \dots, R_4\}$ in the market, for two different runs of the DBE simulation. Both runs had been initialised with the same parameters, for one of them exchange between the SMEs was not permitted, whereas for the other one the SMEs were free to exchange services with one another according to the procedure detailed in Section 4.4.3. In order to train the classifiers used for the exchange decisions, every 500 rounds all SMEs' portfolios were reset to the services they had at round 0. To make comparison easier, the resetting of the portfolios was also done during the run where exchange was not allowed. In effect, in this experiment, "history" repeats itself every 500 rounds. This is the reason spikes occur in the graph every 500 rounds. When exchange is permitted, the SMEs are given the chance to exchange services with one another at rounds 250, 750, 1250, 1750, etc. The graph shows a period of 5000 rounds, when the classifiers have been sufficiently trained.

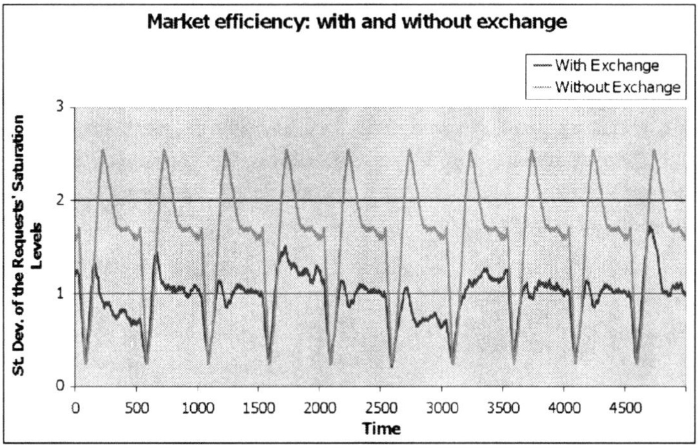

Fig. 4.5: Market Efficiency: We assess the level of market efficiency by plotting the standard deviation of the saturation degrees of the requests in the DBE Market. The smaller the standard deviation, the greater the market efficiency. The graph contrasts these data for a situation in which the SMEs are allowed to exchange services with one another and for a situation where exchange is not allowed. The standard deviation of the saturation degrees of the requests is significantly smaller when exchange is allowed, indicating a more efficient market. For classifier training purposes every 500 rounds all SMEs' portfolios were reset to the services they had at round 0. In the case where service exchanges are allowed, these happen in the middle of each cycle, i.e. at rounds 250, 750, 1250, 1750, etc.

It is evident from the graph, that when exchange of services between SMEs is allowed, the standard deviation of the requests saturation values is considerably smaller. In other words, the requests in the market are more evenly satisfied. This result is quite invariant to initial conditions and parameters of the simulation. So in the system described, not only will SMEs adopt information exchange as beneficial to their individual progress, but it will also result in a global improvement to the efficiency of the market. Again this is in agreement with what is observed in real economies where open standards, publication of innovations and dissemination of ideas lead to highly efficient markets.

4.6 Concluding Remarks

The aim of this work has been to study the rationale as well as the effect of knowledge exchange in economic markets. We focus especially on the software industry; our findings, however, to some extent apply to other industries as well. Sharing of information between commercial firms is considered controversial. Although it is acknowledged that when two companies join forces to develop an innovative product they can both benefit, sharing trade secrets is not undertaken lightly. Our main aim has been to formalise a plausible and elegant explanation of how and why companies adopt information exchange and why it benefits the market as a whole when this happens.

An agent based model of a digital business ecosystem market has been implemented to assist us in understanding the dynamics of the market mechanisms. Firms are modelled as agents with minimal reasoning capabilities. We investigated the properties that emerge from the agent interactions that occur in the market. Specifically, we examined two key characteristics that we observed in the simulations carried out. Namely, the fact that the agents discover themselves that under certain circumstances it is beneficial for them to exchange services and that allowing exchange to take place in the market, makes for greater market efficiency levels.

The technological infrastructure of the DBE will facilitate the dissemination of knowledge among the member SMEs, increasing the volume and the speed of the information flowing in the market. As a result, it is expected that it will allow for greater market efficiency levels in comparison to a conventional market. Admittedly, it is difficult to compare the market efficiency of two different markets. However, an interesting result arose when we performed simulations of the DBE contrasting settings in which exchanges among SMEs were permitted with settings where exchanges were not permitted. Exchanges among SMEs within the DBE further increase the efficiency of the market, which is in agreement with the common intuition that exchanging information is ultimately beneficial for the entire market.

The second and most important conclusion that emerged from the DBE simulation is that exchanges between the agents similar to the ones that happen in real-life arise naturally in our system. At regular time intervals, the SMEs were given the chance to decide whether they wanted to choose a partner and swap some of their services. The decision was taken using classifiers, which were separate for each agent. The agents were not preprogrammed or biased in any way to engage in exchanges. The SMEs, on their own, discovered in which cases exchanging is beneficial for them and what type of partner is the best. Exchange is a practice that emerges, and is not forced upon the agents.

This work does not directly advocate knowledge exchange as a means of increasing profitability of software companies. Knowledge exchange, is indeed an already existing phenomenon in industry as explained in Section 4.3.1. The results presented merely serve as a demonstration of a parsimonious set of assumptions that give rise to exchange in a software market. In other words, we identify the substance of this phenomenon, ridding it from unnecessary assumptions, such as network effects, social issues of trust, or managerial strategies and show the minimal set of assumptions that allow it to emerge.

Chapter 5
Collaborative Query Expansion

Abstract The expansion of the Internet has made the task of searching a crucial one. Internet users, however, have to make a great effort in order to formulate a search query that returns the required results. Many methods have been devised to assist in this task by helping the users modify their query to give better results. In this work we propose an interactive method for query expansion. It is based on two observations: (1) documents are often found to contain terms with high information content which can summarise their subject matter, and (2) some users are more proficient in formulating accurate queries than others. We present experimental results which demonstrate that our approach significantly shortens the time required in order to accomplish a certain task by performing web searches. Furthermore, we propose a scheme under which summaries produced from other searchers' queries are used to propose terms for expansion that lead to reduction of uncertainty about the type of document required.

5.1 Introduction

The growth of the Internet and the increasing availability of online resources have stimulated interest in the field of information retrieval. Information retrieval concentrates on developing algorithms to locate and select documents from a corpus of material that are relevant to a given query. The development of online information retrieval tools, such as search engines many of which utilise hyperlink analysis [Hou and Zhang, 2003], has been greatly beneficial to Internet users. However, many of the users find the current process of searching the web unsatisfactory. This dissatisfaction is not necessarily attributed to the search engine program as much as to the inability of the user to formulate the appropriate query. A user has often only a vague idea of what the relevant query terms may be and has to rely on an iterative process in which the retrieved query results are used to formulate the next query. Interactive query expansion, to which this chapter contributes, is an approach that attempts to assist in this iterative process.

M. Chli and P. De Wilde, *Convergence and Knowledge Processing in Multi-Agent Systems*,
Advanced Information and Knowledge Processing,

We have implemented a system which proposes to the users of a search engine terms that could potentially enhance the search results and lead them more quickly to the target documents. This is done by post-processing the results the search engine produces. Terms from the text of the results are identified based on their information content. This set of terms can serve as a summary of the different types of retrieved documents. In Chapter 4 we illustrated how information exchange between companies can be beneficial both for the companies and for the market. In this chapter we will show how exchanges between individuals in an internet search scenario can make for more efficient search sessions. Summaries produced from other searchers' queries are used to propose terms for expansion that lead to reduction of uncertainty about the type of document required.

Experiments carried out with human users have shown that the terms proposed for expansion by our system, can significantly shorten the session of a web search. These experiments and the results they produced are presented in Section 5.4.

5.1.1 Query Expansion

Research by Spink et al. [2001] has shown that most search engine users typically formulate very short queries of two to three words. Such short queries lack many useful words and do not sufficiently describe the subject that the user wants to search on. In the same work it is suggested that web users tend to go more often from broad to narrow formulation in queries since the most common query modification is to add terms.

The aim of query expansion is, given the user's initial query, to propose possible terms to add to that query so that the quality of the retrieved results is improved.

A number of methods for performing query expansion have been developed. An extensive review on the subject has been produced by [Efthimiadis, 1996].

Early methods involved extracting terms from thesauri which attempted to find more suitable vocabulary as in Gauch and Smith [1991] or performed word sense disambiguation [Voorhees, 1993]. As these proved to be labour-intensive, researchers turned to methods such as lexical co-occurrence [Vechtomova et al., 2003] as well as several forms of clustering [Sparck-Jones, 1971; Leuski, 2001; Loupy et al., 1998; Khan and Khor, 2004]. Lexical co-occurrence is the process of developing relationships between words based upon their co-occurrence in documents. In clustering, documents that share a significant number of terms are grouped together and representative words from each cluster are used for the expansion of the original query. Several algorithms have been proposed for clustering for query expansion. For example, in Loupy et al. [1998] a classification algorithm based on hierarchical clustering is presented, while in Khan and Khor [2004] a hybrid neural network is used to perform the clustering. Most systems, however, that used clustering for query expansion reported rather pessimistic conclusions on their performance [Efthimiadis, 1996]. The similarity of the method proposed here with the methods of lexical co-occurrence and clustering is that the source which provides the candidate terms for the expansion is the set of the retrieved documents as opposed to some knowledge structure, as is the case with the thesaurus-based approaches . As a consequence, if the user chooses terms that do not yield results from the expected domain, the terms a query expansion algorithm will suggest are not likely to be helpful to the user. Coping with this situation without employing strong prior information is too difficult and will not be dealt within this work. See Bodner and Song [1996] for a review of knowledge-based techniques for query expansion.

In this work, it is assumed that the user provides an initial search string which is fairly general and yields results that contain the required class of documents as a subset.

Another method to perform query expansion, perhaps the most effective of all, is that of relevance feedback [Robertson et al., 1986]. It has been used both as interactive [Harman, 1988] as well as an automatic [Mitra et al., 1998] method for expansion. In this method the user submits a query which yields an initial set of results. In the interactive case, the user selects from the set of retrieved documents those he/she believes to be relevant. In the case of automatic query expansion, the system selects the top retrieved documents as the more relevant ones. The system expands the query based upon the terms in the selected documents. Despite the significant improvement in the quality of results this method produces, the research carried out by [Spink et al., 2001] shows low use of the relevance feedback facilities provided in search engines . The low use should not necessarily be attributed to the interactive nature of this method. Sometimes when a user has already found a set of relevant documents they may not wish to expand the query further. Also, relevance feedback algorithms are only useful when relevant documents are returned within the top ranked documents of the results. The method we propose does not have this drawback as it examines a large number of the documents retrieved, much larger than what a human user would realistically be able to examine, to propose the discriminatory terms that will lead to the relevant documents.

More recent methods to perform query expansion involve collaboration of searchers utilising mining user logs [Cui et al., 2003] compiled from the documents the user clicks on as well as constructing user profiles [Nikravesh et al., 2002]. Preliminary results produced by these methods have been promising. The major disadvantage, however, of the methods that rely on implicit user cooperation is the issue of privacy [Cooley et al., 1999]. The method for user collaboration outlined in this chapter, is non-intrusive in the sense that it does not use data associated with the identity of the users. Performed queries are centrally stored, however, without being associated with particular users.

Soft computing methods have recently been used in information retrieval. Examples include fuzzy query [Choi, 2003] and fuzzy grammar [Wang, 2000], which work by attaching weights to the terms of the initial query, as well as the soft semantic web [Martin and Azvine, 2003], a mesh of information linked up in a fuzzy way so that it is easily processable by machines but at the same time efficiently represents the relationships of different notions. Additionally, there has been work that utilises fuzzy association rules [Martin-Bautista et al., 2004] to perform query expansion. Soft computing seems a promising alternative; however the contributions cited above have not yet presented results that compare their performance to those of crisp methods.

Query expansion can be performed manually, interactively or automatically. In interactive and automatic query expansion, the candidate additional terms are provided by the system. The difference between the two methods is that for the interactive method the selection of the terms that are actually added to the query is a decision taken by the user as opposed to the system as is the case for the automatic method. Magennis and van Rijsbergen [1997] and Ruthven [2003] as well as others, have tried to evaluate and compare the efficiency of the two methods. However, in most of the cases, their experiments were based on simulations and not on real human users, with the exception of the "inexperienced" users experiment in Magennis and van Rijsbergen [1997]. Nevertheless the results of the experiments showed that interactive query expansion has the potential to be an effective technique. Even if these results are to be disregarded, the interactive method gives more control to the searcher who knows her utility better than any automated system.

In this chapter we present an interactive method for query expansion. It is founded on the following facts:

1. Documents contain terms with high information content that can summarise them.
2. Terms from queries performed by potentially more experienced searchers can be used to complement a general query string, to optimally reduce the search space for subsequent queries.

We demonstrate query expansion results that testify to the validity of our approach.

The rest of the chapter is laid out as follows: In Section 5.1.2 the rationale behind the method of extracting discriminative terms from the retrieved documents is laid out with Section 5.2 giving details of the metrics used to assess the information content of each word. Subsequently, in Section 5.3 the implementation of the system is described. In Section 5.4 the experiments conducted for the evaluation of the quality of the discriminative terms are reported. The method for collaboration among searchers is described in Section 5.5.

5.1.2 Discriminative Document Terms

Our fundamental observation is that documents contain *discriminative words* that can summarise the document content. This is particularly true for the majority of commercial websites whose authors try to condense the content of the page in small, effective messages. The common characteristic of these discriminative words is the rarity of their occurrence in the corpus of all documents in the database. Borrowing from information-theoretic concepts, we can say that an infrequent word is normally associated with high information content. This implies that such words are normally very good candidates for expanding a query with the exception of two situations. Namely, when the discriminative word is common within the set of the retrieved documents, although it is rare in the corpus, or it is extremely rare in the set of retrieved documents as well as in the corpus. In the first case, that word obviously does not carry any discriminatory power and in the second case the word is likely to have appeared by chance.

A word which falls in one of these two situations is probably not a good candidate for query expansion and should therefore be somehow disqualified. If this intuition is used for expanding the query then these two extremes should be disregarded. In the next section we formalise these intuitions and motivate the term selection model used.

5.2 Term Value

The method proposed is to assign a weight w to each word that appears in the retrieved documents.

Define for a set of documents S and a term t, the set $S[t] \subset S$ consisting of the documents that contain t. Assume we are working within a corpus of documents C. After the user submits the initial query, a set of documents $R \subset C$ is retrieved. Our basic assumption (justified experimentally in Spink et al. [2001]) is that the initial query will be general enough to contain the required document in its results. The goal for the system is then to produce a candidate term for expanding the query. The user is then asked about the relevance of the candidate term t, say, and gives back a "yes"/ "no" answer, essentially selecting between the set of documents which contain t, $R[t]$, and the set of documents which do not contain t, $R - R[t]$. We require a good candidate term to have the following two properties:

1. The expected reduction in uncertainty as a result of the user's choice should be maximised.
2. It must be clear to the user, whether this term is relevant or not with the target document.

The first property can be quantified using entropy. Before the inclusion/exclusion of term t by the user, the required document is one of $|R|$ equiprobable documents. So the associated entropy of this is

$$H_{init} = |R| \times \left(-\frac{1}{|R|} \times \log \frac{1}{|R|} \right) = \log |R| \tag{5.1}$$

When the user is asked to include or exclude documents containing term t, entropy will be reduced to either $\log |R[t]|$or $\log |R - R[t]|$. If the two possible replies are considered equiprobable, the mean reduction in entropy caused by t will be

$$\Delta H_t = \log \frac{|R|}{|R[t]|^{1/2} \times |R - R[t]|^{1/2}}. \tag{5.2}$$

The second property is harder to measure theoretically and yet it is a crucial part of the system. Used on its own, the first property will produce a multitude of candidate terms which may provide good uncertainty reduction but which may not be helpful to the user. The heuristic we apply here is the specificity part of the traditional TF.IDF scheme (IDF term) which is given by

$$IDF_t = \log \frac{|C|}{|C[t]|}. \tag{5.3}$$

This measure is known in work by Salton and Buckley [1988] using TF.IDF to favour rare, information-rich terms. In our case it acts as an artificial negative bias that will eliminate possible candidate terms that happen to satisfy the first requirement purely by chance. A potential problem of this approach is that it may also bias candidates towards rare, over-technical and potentially unfamiliar terms for the user. In practice, for the document sets tested in this work this effect is minimal.

The two terms described above are combined into a single term, to which we refer as weight w:

$$w(t) = \Delta H_t \times IDF_t. \tag{5.4}$$

Another useful measure is that of the effectiveness of a query. For a query q and its associated set of high-information content terms $\mathbf{t}$, effectiveness e is defined as follows:

$$e(q) = \frac{1}{n} \sum_{i=1}^{n} w(t_i), \tag{5.5}$$

An effective query is one which yields a homogeneous set of retrieved documents. As a result, the set of high-information content terms $\mathbf{t}$ associated with it will have small weights w, i.e. low discriminative power.

5.3 Implementation

In summary the algorithm for extracting discriminative document terms proceeds as follows:

1. The user submits a query to a search engine.
2. The first K results the search engine returned are collected and parsed.
3. Stemming is performed on every term that appears in the collection.
4. The value of every term is calculated, as discussed in Section 5.2.
5. The terms are sorted according to their cost and a list of the first N of them is presented to the user, as candidate terms for expansion.

In the rest of this section we discuss the algorithm in more detail.

5.3.1 Initial Phase

The procedure is started by the user typing in a query. This query is then executed by a search engine. Since our method is applied during a post-processing phase, it can be used with any information retrieval system which returns a list of documents. The Google search engine was used during the experiments we carried out. Stemming and stop word removal on the query were left to the search engine to do. Subsequently, the first K result pages returned by the search engine are read in and parsed, in order to extract from them their actual textual content lemma by lemma. For the experiments described in Section 5.4, 200 result pages were read in each time. Multiple occurrences of a lemma within the same document are not taken into account.

5.3.2 Stemming

The next step is to perform stemming on the terms of the retrieved documents. A stemming algorithm tries to reduce a word variant to its root form. The root is the form of a word after all affixes have been removed. We have implemented a form of the KSTEM stemmer proposed by Krovetz [1993] which works by removing suffixes from a word variant piece by piece and after each removal looks up the reduced word in the dictionary. In our version we use WordNet, a lexical database for the English language [Miller, 1995]. Krovetz originally used LDOCE, the *Longman's Dictionary of Contemporary English* . The basic idea is that, if a word is in the dictionary, this means that it has a different meaning from its root and it should not be stemmed further. For example the verb "fought" will be reduced to "fight" and the noun "fairies" to "fairy", however it will not be further reduced to the word "fair". Additionally, the word "cooked" will not be reduced to "cook" as WordNet recognises it in its initial form as an adjective. KSTEM has been found to improve retrieval effectiveness over the well-known Porter stemmer [Porter, 1980] which does not stop stemming until it reaches the root of a word [Sanderson, 1999].

This practice of reducing all the encountered lemmas to their root form ensures that terms with the same stem, meaning and information content such as the words "memorise", "memorised" and "memorising" will not be counted twice in the same document.

5.3.3 Common Word Filtering

As laid out in Section 5.2, where the term value calculation is described, an important parameter is the information quantity of a word; a measure of its rarity in the corpus of all documents. An alternative, simpler approach would be to employ a stop list of very common words. While this was also seen to produce good results, using a database of term frequencies in the corpus and calculating the information quantity of each encountered term is a more elegant approach that allows for smooth filtering. A term that is quite commonly used in the corpus, but is used even more frequently in the collection of retrieved documents is potentially one of high significance in the domain that is being searched. It is also likely to be beneficial if used for the expansion of the initial query. A stop list would eliminate such a word, whereas the use of a database allows it to proceed.

In the initial stages of the implementation, the frequency statistics database used was the Brown Corpora list of 2000 most common English words [Francis and Kucera, 1979]. This list has been compiled from a large number of English literary texts. Although it did filter out numerous of the most common words of the English language such as "the" and "a", it allowed words such as "link" and "email" to occupy the top of the lists of words suggested for query expansion for most of the experiments carried out. As one would probably expect, the vocabulary used in English texts found on the Internet is somewhat different to what is used in the literary texts. In order to overcome this problem, a list of common words encountered on the Internet, similar to Brown's had to be specially compiled from English web pages. The employment of this list as the frequency statistics database of the system greatly improved the results of the experiments performed.

5.3.4 Term Selection

As soon as the user has entered his or her query, the results are returned from the search engine and all the encountered lemmas are read in; then the score is calculated for each word in the retrieved documents, as described in Section 5.2.

The user is subsequently presented with the candidate words clustered according to the documents they appear in. The user selects the lemmas she sees fit, which are then added to the original query with the connectives AND, or AND NOT.

This process is repeatedly applied to any subsequent set of results.

5.4 Evaluation

In order to evaluate the quality of the discriminative terms extracted, a set of experiments has been carried out. Human users have been asked to perform a series of tasks by carrying out web searches. The Google search engine has been compared to the system we propose and a significant decrease in the length of the web search sessions has been recorded. This section describes the experiments in detail and lays out the results they produced.

5.4.1 Experiments

As stated in Section 5.1.1, the algorithm proposed operates under the assumption that the initial search string that is provided by the user of the system is fairly general and yields results that contain the required class of documents as a subset. This assumption is supported by research carried out by Spink et al. [2001], which suggests that web users tend to go more often from broad to narrow formulation in queries since the most common query modification is to add terms. The method we propose will process the retrieved documents and will propose to the user a set of words that in a way summarise the types of retrieved documents. At this stage, the users are asked to use these terms to expand their initial query, narrow down the retrieved documents and eventually, after a number of iterations, isolate the required subset of results.

One way to evaluate the system is to test the ability of the proposed algorithm to identify discriminative words which reflect the existence of discrete subsets of documents encapsulated in the collection of retrieved results. A method to do this is to form queries which can be semantically interpreted in more than one way. Such queries are expected to yield several groups of documents, each group corresponding to a different sense of the query. It will be interesting to see whether the implemented system actually suggests words from the retrieved documents that reflect those distinct meanings.

For example, when our users were given the task, "Find two websites where several British customs are described" many of them formed the query, "British customs". This query can be interpreted in at least two ways. The word "customs" may mean a "specific practice of long standing", or it may mean "money collected under a tariff" [Miller, 1995]. Performing a trial search on Google one may see results related to HM Customs and Excise web page, news articles that report recent activities HM Customs have been involved with, web pages that describe British traditions, etc. When the above query is given as input to the implemented system and the first 200 results that Google returns are read in and processed, the first 30 suggestions the user is presented with are the following: tradition, excise, include, carry, issue, create, HM, comment, party, condition, medium, regard, cigarette, department, duty, thousand, organisation, follow, source, parliament, soldier, country, investigation, serve, traveller, citizen, law, drug, nation, vehicle.

The word "tradition" is included in the list representing one of the interpretations of the query as well as the words "excise", and "HM" which are associated with the second interpretation. The user selects the word "tradition". This time the user does not need to incorporate any more terms to the query as the search engine returns documents that are of interest to the user on the top of the first page of results.

In the experiment described above, only two iterations were enough to prune the enormous amount of retrieved documents and to lead the user to the documents she was looking for. Research carried out by Spink et al. [2001] suggests that the mean and median number of queries per user session were 4.86 and 8, respectively. If we assume that during a session a user enters an initial query, observes the results, then modifies the original query and tries again, this statistic shows that the user has to do quite a few query modifications before they are presented with the required documents.

5.4.1.1 Experimental Setting

The evaluation of the system was not based solely on ambiguous queries. For a wider assessment we used the tasks defined for the Interactive Track of TREC-10 [Hersh and Over, 2001; White et al., 2001] plus some others defined by us. The searches were run on the Internet instead of a precompiled collection, to provide for a more realistic setting.

In total, 24 subjects participated in our experiments. All subjects were educated to graduate level and were recruited from various departments of the University of Cambridge, including the Departments of Engineering, Classics, Computer Science, Physics, Biological Sciences and Medicine. The average age of the subjects was 27.06 with a standard deviation of 5.27 years. All users used computers and the Internet frequently as part of their work. The average experience of online searching among the subjects was 6.38 years. All users cited Google as their favourite search engine. Native English speakers were 40% of the users. The following are the tasks used in the experiments.

1. Find a website likely to contain reliable information on the effect of second-hand smoke.
2. Tell me three categories of people who should or should not get a flu shot and why.
3. List two of the generally recommended treatments for stomach ulcers.
4. Identify two pros or cons of taking large doses of vitamin A.
5. Get two price quotes for a new digital camera (3 or more megapixels and 2x or more zoom).
6. Find two websites that allow people to buy soy milk online.
7. Name three features to consider in buying a new yacht.
8. Find two websites that will let me buy a personal CD player online.
9. I want to visit Antarctica. Find a website with information on organized tours/trips there.
10. Identify three interesting things to do during a weekend in Kyoto, Japan.
11. Identify three interesting places to visit in Turkey.
12. I'd like to go on a sailing vacation in Australia, but I don't know how to sail. Tell me where I can get some information about organized sailing cruises in that area.
13. Find three articles that a high school student could use in writing a report on the *Titanic*.
14. Tell me the name of a website where I can find material on global warming.
15. Find three different information sources that may be useful to a high school student in writing a biography of M. Jordan.
16. Find a book with lots of information for a high school report on the history of London.
17. Find two websites where several British customs are described.

18. Identify two rules of American and two of European football.
19. Name one of the dangers jaguars face.
20. Name a bond considered among the strongest in 2002.

Each user was asked to complete a set of ten tasks selected randomly from the list above. These were in turn randomly divided in two groups of five tasks, one group to be completed using Google and the other using Google and query expansion terms suggested by the proposed system. A researcher observed each user while carrying out their assigned tasks and noted the queries they used, the iterations it took them to complete each task and the number of documents they viewed. The users were allowed to read the summaries of the documents Google provides, but only the sites they clicked on counted towards the documents viewed.

5.4.2 Evaluation Results

As can be seen in Table 5.1 the proposed system diminishes significantly the average number of iterations and documents per user search session in comparison to Google.

	Google	Proposed System
Average iterations	1.9375	1.425
Average documents viewed	4.2	1.8625

Table 5.1: Average iterations and documents viewed per user session.

The fact that Spink et al. [2001] recorded much longer user sessions in their work (the mean and median number of queries per user session were 4.86 and 8, respectively) for sessions carried out using Excite than we recorded for tasks carried out on Google may be due to several reasons. Apart from the search engine used being different, the users in our experiments were all of a graduate level of education and had quite some experience in web searching. In addition, the tasks the users were given were fairly well specified. In many cases users search without having in mind a complete specification of the problem they are trying to solve.

In Tables 5.2 and 5.3 some of the queries used by users for the execution of the given tasks are presented as well as the first 30 words that the system suggested for expansion of each one of the queries. These words give an overview of the set of retrieved documents without the user having to look at the documents one by one and make for shorter user sessions.

We use a variant of k-means clustering in order to group the retrieved words according to the documents they appear in together. In this technique, instead of a mean of a cluster of terms, we used the term in the cluster with the minimum average distance to other terms in the cluster. The distance between two terms was given by the number of documents containing both terms. In the two tables the clusters are shown in different rows.

Initial Query	Suggested Terms for Expansion
British customs	duty, excise, issue, create, HM, comment, condition, media, cigarette, vehicle, department, carry, thousand, organisation, source, soldier, drug, law, investigation, traveller, citizen
	serve, tradition, include, party, follow parliament, country, regard, nation
book London	booking, offer, reservation, attraction, discount, airport, deal, travel, budget, rooms, facility, rooms, facility, luxury, tour, ticket, park, secure, star, city, night, accommodation, garden, shop, stay, apartment, restaurant, street, map, location
	ISBN, publisher
football rules	winner, overtime, scores, tie, played, interception, playoff, playing, guard, kickoff, offence, consist, pick, scoring, touch, completion, quarterback, tied, punt, roster, touchdown, defensive, injury, fumble, sport, defence, foot, coin, regulation,
	soccer

Table 5.2: Tested queries and suggested expansion terms.

A few interesting aspects about some of the results are pointed out below. **"book London" query:** The results are dominated by sites which advertise their services on hotel and flight bookings and reservations. The terms "ISBN" and "publisher" though, are an indication that a few sites on books on London have been retrieved as well.

"football rules" query: The list of results is flooded with football terminology and this was to be expected; however, it is interesting to see the word "soccer" included.

"jaguar" query: The results are mostly populated by the most commercial of the meanings, that of the car. Nevertheless the presence of the terms "cat", "tail" and "panther" depicts the existence of a number of websites that refer to the animal.

"bond 2002" query: The query yields an interesting mix of results. This is due to the multiple meanings of the word bond. The majority of the keywords in the

Initial Query	Suggested Terms for Expansion
jaguar	car, feature, picture, photo, club, logo, engine, frame, racing, enthusiast, join, drive, tip, sport, event, classic, rally, driver, jag, model, tail, forum, item, driving, preview, video, speed
	cat, tail, panther
bond 2002	tv, villain, actor, gadget, EON, Thanksgiving, soundtrack, upload, autograph, spy, poster, Jinx, trailer, boxed, instal, james, marathon, baddie, stunt, convertible
	investment, coupon, portfolio, treasuries headline, yield, upgrade, swap, dividend,
	chemical
yacht	ft, meters, frame, schooner, specification wooden, swan, sabre, boatbuilder, keelboat, creek, craft, motor,sailing
	club, cruise,sail, racing, regatta, bay, event, marina, fleet, sailor,ocean charter, marine, crew, sea, championship,

Table 5.3: Tested queries and suggested expansion terms (cont'd).

list is related to the James Bond film: villain, actor, gadget, EON (the producing company), Jinx (a character in the film), convertible (referring to a car in the film), etc. The words "Thanksgiving" and "marathon", although looking unrelated to the subject, refer to a James Bond film broadcasting marathon to be screened by a U.S. television network on Thanksgiving day, an event which as is apparent from the retrieved documents, had been heavily advertised on websites related to the film. One of the suggested words is associated with chemical bonds. The rest of the results reflect the financial meaning of the term "bond": investment, coupon (as in coupon rate), treasuries, yield, swap and dividend. The word "headline" is very common in sites which refer to financial bonds and this is the reason it appears in the top 30 suggestions.

"yacht" query: The terms that the system proposes to the users for expanding the query, in a way answer the question that was given in the task: "Name three features to consider in buying a new yacht." One of the clusters is populated by terms which are either features of yachts or different types and brands of them. As it is apparent from the contents of the other cluster, there is a group of websites retrieved that concern regattas and racing.

5.5 Introducing User Collaboration for Query Expansion

A fundamental observation that motivated this work is that some users are more proficient in formulating effective query strings than others. This may be attributed to their background, their expertise in the subject being queried or their experience in searching. Terms from queries performed by experienced searchers can be used to complement a query string formulated by less precise searchers, to optimally reduce the search space and lead to the desired document faster.

The algorithm set out in Section 5.2, returns as output a set of terms with high information content. This set acts as a summary of the types of retrieved documents. As shown in Tables 5.2 and 5.3 the terms are clustered according to collocation in documents. The clustering often reveals different groups of documents in the retrieved set. For example, there were two clusters in the set generated from the query "British customs" reflecting the dual meaning of the term "customs". An experienced user would have perhaps used the query "British traditions" in order to solve the task "Find two websites where several British customs are described." The term "tradition" is shortlisted as a term with high information content; however, it stands among other less discriminative terms.

While this method produces intuitive results, it relies upon the user specifying the number of clusters which is inconvenient in practice. The collaborative method proposed in this section, employs effective query strings used by other searchers in the past to give the high-information content terms shortlisted a meaningful ordering.

5.5.1 Collaboration Procedure

The method of collaboration proposed in this section uses the database of stored queries to rank the discriminative terms produced by the current query. More specifically, a discriminative term's rank improves according to (a) the number of effective stored queries it appears in, and (b) the number of terms in these queries that appear in the set of discriminative terms.

The algorithm used for the collaborative query expansion is as follows:

1. Every time a user performs a query the system generates a list of the 30 terms with the highest information content extracted from the set of retrieved documents as described in Section 5.2.
2. The effectiveness of the query is calculated as per Equation (5.5).
3. The set of discriminative terms of the current query is compared to each of the queries stored in the database. The method used for this comparison is described in Section 5.5.2 below.
4. Each stored query is assigned a score according to its effectiveness and to the number of terms it has in common with the set of discriminative terms.

5. The high information content terms associated with the current query are ranked according to the score of the stored queries they appear in. Subsequently, the best of them are presented to the user as candidates for expansion.
6. The current query is stored centrally in the database along with its effectiveness value.

5.5.2 Comparing Sets of Terms

Assume the database holds m stored queries. Each stored query $\mathbf{q_i}$ in the database is associated with a value of effectiveness e_i. This is calculated using Equation (5.5) ($e_i = \frac{1}{n}\sum_{i=1}^{n} w(t_i)$). Further assume that the current query yields a set of discriminative terms $\mathbf{t}$.

Define $\mathbf{q_i} = (\alpha_1, \alpha_2, \ldots, \alpha_n)$ and $\mathbf{t} = (\beta_1, \beta_2, \ldots, \beta_n)$, where n is the number of terms in the corpus and $\alpha_k, \beta_k \in \{0, 1\}$ depending on whether the kth term is contained in the vector.

The match between $\mathbf{q_i}$ and $\mathbf{t}$ is defined as the similarity:

$$sim(\mathbf{q_i}, \mathbf{t}) = \mathbf{q_i^T} V \mathbf{t}, \tag{5.6}$$

where V is the matrix:

$$V = \begin{bmatrix} v_1 & 0 & 0 & 0 \\ 0 & v_2 & 0 & 0 \\ 0 & 0 & \ddots & 0 \\ 0 & 0 & 0 & v_n \end{bmatrix} \tag{5.7}$$

and v_k is the IDF weighting of the term k, $v_k = \log(\frac{|C|}{|C_k|})$, as given in Equation (5.3).

The score s_i of each stored query q_i is calculated as follows:

$$s_i = e_i \times sim(\mathbf{q_i}, \mathbf{t}). \tag{5.8}$$

Subsequently terms in $\mathbf{t}$ are assigned a score u_j according to the score of the stored queries they appear in.

$$u_j = \sum_{\forall i,\, t_j \in q_i} s_i. \tag{5.9}$$

The top scoring stored terms in $\mathbf{t}$ are shortlisted and presented to the user as candidates for expansion.

5.5.3 Example of Collaboration

For the collaboration to be effective, multiple participants with intersecting goals should be involved. The more queries are stored in the database the greater the precision of the proposed expansion terms. In this section we give an example of how the task "Name a bond considered among the strongest in 2002." can be effectively solved with the aid of the collaborative query expansion method proposed.

The user enters the query "bond 2002" and the list of discriminative words shown in Table 5.3 is generated. The discriminative terms are shown in the second column of Table 5.4, ordered according to their discriminative power w [Eq. (5.4)]. This query is not effective enough as it produces results related to three different areas: the James Bond films, the financial sector and chemistry.

Each stored query is compared to the set of discriminative terms. As an indication, the five top ranked stored queries are the following: "James Bond", "James Bond trailer", "chemical bonds tutorial", "Bond portfolio", "treasury bond dividends". The third column of Table 5.4 shows the discriminative terms ordered according to the score u [Eq. (5.9)].

It can be seen in Table 5.4 the terms that appear in high scoring stored queries (e.g. "James", "trailer", "chemical", "portfolio", "treasury", "dividends") occupy the top of the list when the score u is used to rank the discriminative terms.

5.6 Limitations and Future Work

As with any query expansion method that uses the set of retrieved documents as its source from which to extract candidate terms for expansion, this method requires that the user inputs an initial query which is reasonably broad. If the set of retrieved results does not contain as a subset the documents the user is interested in, the words that will be suggested will not be useful for query expansion. However, they may still serve as a summary of the retrieved results and help the user to reformulate the original query.

Another potential limitation arises from the fact that a significant amount and diversity of stored queries is required to produce recommendations which are beneficial. In other words, for the collaboration to be effective multiple participants with various but intersecting goals should be involved.

Future work plans include improving the collaboration protocol such that the time and computational power required for post-processing is minimised.

Another avenue of future work is to make it possible for the proposed terms for expansion to be presented to the user in the form of a semantic web instead of a ranked list. It is stated in Han et al. [2006] that just as the ranking of documents is a critical component of today's search engines , the ranking of relationships will be essential for tomorrow's search engines that will potentially support discovery and mining of the semantic web. However, ranking a set of interconnected entities and relations is more complex than ranking a set of documents or paths of semantic

Rank	Before collaboration ordered by w	After collaboration ordered by u
1	tv	James
2	investment	trailer
3	villain	chemical
4	actor	portfolio
5	coupon	treasury
6	gadget	dividends
7	eon	actor
8	thanksgiving	spy
9	soundtrack	yield
10	upload	jinx
11	portfolio	dividend
12	autograph	swap
13	treasuries	villain
14	spy	soundtrack
15	poster	coupon
16	headline	stunt
17	jinx	autograph
18	yield	gadget
19	chemical	upgrade
20	trailer	tv
21	boxed	convertible
22	instal	poster
23	upgrade	headline
24	swap	instal
25	marathon	eon
26	baddie	upload
27	dividend	marathon
28	stunt	boxed
29	convertible	thanksgiving
30	James	baddie

Table 5.4: Order of discriminative terms for the query "bond 2002", before and after the collaborative query expansion algorithm was applied. Note how the terms that appear in high scoring stored queries (e.g. "James", "trailer", "chemical", "portfolio", "treasury", "dividends") occupy the top of the list when the score u is used to rank the discriminative terms.

associations. The semantic ranking approach considers the total number of entities and relations that match a user's interests by assigning a value of calculation to each of them. How to compute the semantic similarity is a critical issue in semantic ranking . We plan to expand our work by developing an approach that will take into consideration such criteria as the similarity between two entities and their similarity reflected in context in terms of the documents they appear in. User-friendliness can be improved by building a network of nodes that contain information. The nodes will be connected by links that indicate similarity. The similarity measure is based on co-occurrence of terms in documents.

5.7 Conclusions

In this chapter we have presented an interactive method for query expansion based on subdivision of retrieved documents as well as user collaboration. The method is founded on two facts: (a) the expertise of some searchers might be useful to others, and (b) documents contain some terms with high information content which can summarise their subject matter.

High-information content terms are extracted from the collection of the retrieved results. The information quantity they carry as well as their ability to prune down the search space, is then assessed. To evaluate the quality of terms produced, experiments were carried out with human users. The users were given a set of tasks to complete, using a mainstream search engine. Top-ranking words extracted from the retrieved documents were presented to the users as a list of candidate terms for expansion. Tasks carried out on our system were compared with others carried out on Google alone. In terms of both the number of iterations and the number of documents viewed until each task was completed, the user sessions recorded on the proposed system were significantly shorter.

The collaborative query expansion method proposed further improves the performance of the expansion term recommendation mechanism. Using a database of stored queries and their associated list of high information content terms it suggests suitable terms for query modification, in the same way a more experienced searcher would assist a novice one in formulating an precise query.

To our knowledge, this method is the only collaborative technique that is non-intrusive. In other words, unlike other collaborative methods, such as profiling or user log mining, it does not invade the privacy of the users by recording data that is associated with their identity and preferences. Instead it records the query strings the users enter, a practice already common among search engines.

Chapter 6
Micro-economic Control of Distributed Intelligent Personal Assistants

Abstract In this chapter, we describe a stable strategy for a realistic large-scale distributed system. This system consists of automatic personal assistants (PAs) that can book time slots in each other's diaries, and that have to pay for doing so. We analyse stable strategies for PAs in the light of the definitions in Chapter 3. We develop a computational model of the resource management tasks of a general PA. The scheduling of time slots is crucial in this. By assigning values to resources, we propose a micro-economic trading strategy for the PAs. This is a free market strategy with inflation and discount on bankruptcy. We simulate the network of PAs using this strategy, with a random stream of incoming requests for time slots. We find that the strategy is stable and robust.

6.1 Stable Strategies

Stability of strategies was first studied in an evolutionary context [Hines, 1987]. We cannot use this concept, because our network of personal assistants does not evolve as a population, the number of PAs is fixed. We have dealt with a varying number of agents in Section 3.3.2.1, but this is a level of sophistication that we will not need in this chapter. A related notion of stable strategies arises in the context of dynamic or differential games [Olafsson, 1995]. If we wanted to use these games for modeling the evolution in time of the diaries of the PAs, the resulting dynamic system would contain time delays due to the forward booking of time slots, and this would make it very difficult to analyze the stable solutions. The stability of nodes with a strategy can also be seen as a control problem [Åström and Wittenmark, 1995]. In this case, the set of individual strategies is the control action, and the function to be optimized is some positive definite combination of how far the individual nodes are from a local optimum. We have not adopted this stable control because we do not think that is practical to derive it for the network of PAs.

Instead, we choose an economic approach, using market mechanisms for allocating the time slots of the PAs. The PAs are software agents in an information

M. Chli and P. De Wilde, *Convergence and Knowledge Processing in Multi-Agent Systems*, Advanced Information and Knowledge Processing,

economy, and engage in dynamic pricing of their services, as in Kephart et al. [2000]. This allows the PAs to buy and sell time, and to adapt their prices. This can be called *micro-economic control of distributed systems*, and was developed in Zlotkin and Rosenschein [1996]; Rosenschein and Zlotkin [1994] and [Sairamesh and Kephart, 2000; Tesauro and Kephart, 2000]. The strategies of our PAs are parameterized, i.e. every PA has the same strategy, only differing in the values of the parameters. We call the strategy stable if its prevents the network from "freezing," where none of the PAs buys or sells anymore. A stable strategy is robust if substantial noise on the parameters and the initial conditions maintains stability. In this chapter, we investigate the robustness of one strategy, via simulations.

The resources that the PAs buy and sell are the time slots they own. The time slots, their prices, and the cash a PA owns, together constitute the state of a PA. (We will talk of value rather than price in the rest of this chapter. Value is numerical, but can include more than monetary value.). This state is a random variable, because the requests for time slots are generated stochastically. This causes randomness in the trades of the time slots, and ultimately in the state of the PAs.

We will only consider trading strategies that are based on the current state of the PA. This differs from trading on the stock market, for example. A trader on the stock market may take into account the history of his or her successful trades, when making a trade. As our trading strategies do not take past states into account, and are stochastic, the resulting evolution of the state of the whole system of traders has the Markov property. This is an approach that we have taken throughout this book. There are many reasons why an agent may want to ignore the past. Speed of decision making is one factor . Multi-agent systems tend to be complex, and the states of the other agents provide a lot of information already. Moreover, if a market is efficient in terms of the efficient market hypothesis [Elton et al., 2006], then all information is included in the present state, it is not possible to learn from the past. Another reason for ignoring the past is that there is always a bias in analysing the past. What time window to use? Discount events that occurred a long time ago? Apply smoothing?

Once we accept the Markov property, we can use the definition of stability from Theorem 3.2: the system of trading PAs is stable if the system converges to an equilibrium distribution. In this chapter, we add the condition of no "freezing," i.e. at least some PAs have to be able to continue trading time slots. This is merely ruling out trivial equilibrium distributions, where all states have a probability of either 0 or 1. In this chapter, we also add the notion of robustness, which is stability under perturbation of the parameters. We will investigate robustness via simulations in this chapter because we lack the explicit knowledge of the transition matrix of the Markov chain to derive formal perturbation results for the system of PAs.

Micro-economic approaches are not the only successful ones for scheduling . Fuzzy logic has been successfully applied to scheduling [Lee and Pan, 2004; Martin and Azvine, 2003]. It captures aspects of how humans can exploit uncertainty in reaching an agreement about time slots. Fuzzy methods can be implemented on top of the system we propose here. Agents have also been used extensively for scheduling [Zhang et al., 2006; Maldonado et al., 2005]. Agents are good at modeling negotiation strategies, and rationality (such as in BDI agents). Agent-based approaches, because they are good at modeling deviations of rationality, are often less optimal than micro-economic approaches.

6.2 Network of Intelligent Personal Assistants

In the sequel, PA or automatic PA refers to a computer. The term "real PA" refers to a human being. A PA's user can be a real PA or the boss of a real PA. Many statements apply to automatic PAs as well as real PAs.

6.2.1 Definition of the Automatic PA

Intelligent time management is more than keeping a diary. It is necessary to be flexible, and to be able to reschedule the diary. Extra time may need to be made available, by borrowing time allocated to other activities. The owner of the diary will have a preference for a particular time allocation. This can be modeled by attaching a value to the time, so that an allocation has a value, and that the allocation with the lowest value or cost can be preferred. The value of time may depend, not only on the moment of the day, but also on other allocations that have been made. Five minutes between two meetings will have another value than the same five minutes on an empty day.

In scheduling time, the physical location has to be taken into account. Two contiguous time slots should have the person in the same or in adjacent locations. A system that schedules time slots, with a value and location attached, can be used to schedule any resources. This can be money, investments, goods, a store, etc. It can even be other personnel, with their own diary and time management.

A PA also gathers information. Whereas time scheduling gives information about what will happen at some future point in time, the information that a PA collects is about activities that have happened at some past moment in time. *This symmetry between past and future will be essential in our computational model of an intelligent PA*. An activity is first scheduled, it is in the future. Then it becomes current, and while it is taking place, it is documented or described. Once the activity has taken place, its description becomes information about a past event. As an example, consider a meeting. At first, it is scheduled in a diary. The diary contains an agenda. Then, the meeting is held, and minutes are taken. The minutes of the meeting form a record that can be inspected later.

The basic structure for the intelligent PA will be, for every resource, its location in time, its value as a function of time, and its description as a function of time. If there are n resources, we will denote by $l_i(t)$ the location of resource i at time t, by $v_i(t)$ the value of resource i at time t, and by $d_i(t)$ its description. We assume that the locations are numbered, so that $l_i(t)$ is a number. The value $v_i(t)$ is de facto a number. The description $d_i(t)$ is a file that may contain text and drawings. We will discuss later why d does not include voice.

Consider again the example of a meeting. Employee i has to attend the meeting between the times t_1 and t_2. Say the room number is location 3. Then

$$l_i(t) = 3, \quad t_1 < t < t_2.$$

The employee knows that the first five minutes of the meeting are seldom interesting, because some participants usually are late. The value function could be (we assume time is measured in minutes)

$$v_i(t) = \begin{cases} 0, & t_1 < t < t_1 + 5, \\ 4, & t_1 + 5 < t < t_2, \end{cases}$$

indicating the employee values his time at four units per minute during the effective part of the meeting.

The function l could change, if it is decided to change the location of the meeting, or the start or finishing time. The function v can change if, for example, it becomes known that somebody will participate who is usually very late, and without whom the meeting cannot start.

The meeting is in the future for $t < t_1$, and d will then contain only the agenda. During the meeting, minutes are taken, and directly entered into a file. For times $t > t_2$, the description d contains the minutes of the meeting.

6.2.2 Further Specifications

Existing project planning software is often perceived as inflexible by its users, because the predefined time slots are not appropriate. Our model avoids this by defining time slots so small, that time becomes virtually continuous. One-minute time slots are motivated by the observation that some manager's interactions are as short as 2 minutes.

The near-continuous time has other advantages. Priority is modeled by attaching a value to a resource . This value is a function of time, just as the location. By making the time increments small, the value can change independently from the location. This was illustrated in the example where the value of the first 5 minutes of a meeting was smaller than the value for the rest of the meeting.

The reason why we use value instead of priority is that priority does not make sense for past times. The value of a resource in the future is its estimated or budgeted price. The value in the past is the price the resource had when it was current. For example: the value of a specific workstation in 2 years' time may be estimated at £2000 (actual price in 2 years' time). Value of a specific workstation 2 years ago was £4000. If t is the current time, this means $v(t + 2yr) = 2000$, $v(t - 2yr) = 4000$.

The description consists of ASCII text and graphics, but not voice. This is because d has to be archived, and also exchanged with other nodes. We assume a record is kept of every voice contact. For a computer unfamiliar with the originator's voice, it would be impossible to search for a particular content, or word, if the speech was not transcribed into text. Therefore we will assume that every voice contact is transcribed into text, either by a speech recognition system, or simply by the person talking, taking written notes on a keyboard, as journalists conduct interviews nowadays. Sketches, or whole pictures can be added to this. Later, when d has to be searched, the text can be searched for occurrence of keywords, and the pictures can be searched with pattern matching techniques. The ability to retrieve the record d is important, because a large part of a PA's work consists in research of information. Putting information into d, even if it can't be automated right now because of limitations of speech recognition, has to become a way of (intelligent) being.

Transcription into text at the PA's site also allows the use of a speech recognition system fine-tuned to one user. A real PA talks 75% of the time, so it is important to have fast speech recognition. Text coding is very efficient for speech. A whole year's talking of a real PA would only take about 50 megabytes storage, if coded in text.

A PA's work consists in scheduling resources and finding out information. Both tasks require communication with other PAs. This means that the locations, values and descriptions have to be available. Giving read or write access to certain other PAs can do this. Thus, for every $l(t)$, $v(t)$, and $d(t)$, there will be a list of PAs that have read access or write access. We assume that write access implies read access. For example, your PA will give the PA of the organizer of a meeting write access for the location of that meeting. If somebody important organizes the meeting, you may allow the organizing PA to change the time of the meeting, but not in a weekend, for example. The organizing PA would then have write access for $l(t)$, for values of t outside the weekend.

In order to regulate the access of other PAs or their users, they have to be known. If the other PA is an automatic one, or a real one on the phone or a video link, this just means caller identification. If an employee wants to talk in the office to another employee who has an automatic PA, an electronic badge could signal the identity of the visiting employee to the PA. Moreover, in such a situation, the PA would use speech recognition to note down what is being discussed, or one of the two employees would type a summary of the discussion into a computer.

6.2.3 The Intelligent Automatic PA

We will compare here the intelligent PA with the different notions of intelligence. Psychometric tests caused us to include pattern recognition in an intelligent system. In the intelligent PA, pattern recognition is used while searching d. This can be matching of keywords, or approximate matching of figures.

Factor models are used to distinguish several kinds of intelligence. They are not applicable here because the intelligent PA is a specialized piece of software, with data structures adapted to the work of a PA. A PA can only be intelligent as a PA, and not as anything else.

According to Piaget [1964], the intelligent being makes a special use of structures. So does our PA. The structures are the functions l, v, and d, and the access lists. We have described before how these functions are defined and modified by several PAs. The information is contained in the times, locations, access lists, values, and description of resources.

From Piaget we now borrow the notion of *creation of new structures*. Assume that in a meeting, it is decided to hire a new member of staff, or buy a new piece of equipment. At that moment, new data structures l, v, and d are created for the new entity. The structures are not only created, their existence in time varies as well. Ten years' time may be much too far ahead to predict anything about certain entities. Take for example the computer you are currently using. Will you need to know its location and value in 10 years' time? If not, it's l, v, and d (the state it's in, the data it still contains), will be undefined for values of time 10 years ahead. For the same reason you may want to purge data that is old. Its description becomes then undefined. It is now clear that the data structures in our intelligent PA can be created and destroyed. They can also become temporarily undefined. They have a dynamics very close to the dynamics of Piaget's structures.

The intelligent PA adapts to its environment. It is open to changes. Time, location, and description change as circumstances change. They can also be altered by a restricted number of other PAs.

Another way in which project planning software is seen as inflexible is that it does not take into account that other partners in the project have their own planning. This is our reason for considering a network of PAs. Only in the network can intelligence emerge. A single PA, isolated from its environment of other PAs, cannot function.

6.2.4 Negotiating and Optimizing Agents

A diary is filled with commitments that bind the owner of the diary. The commitments are made with respect to a network of other diaries or PAs. If one PA alters its functions l, v, or d, the functions of other PAs may have to be altered as well. *This makes a network of PAs into a dynamical system.*

This shows the need for a negotiating agent in a PA. If another PA wants to change, for example, the time of a meeting, the negotiating agent will check whether this is possible, and either accept, or issue another proposal for a time change. If all goes well, the agents will settle for a mutually agreeable time.

This negotiating process has to be refined. The fact that a value is attached to time allows us to model the borrowing of time. In this model, the PA starts with a total amount of money or units of value, and will try to make this amount as large as possible over time. Participating in a meeting will earn money. Organizing a meeting will cost money, to be paid to the participants. Buying material or hiring somebody to work for you costs money. The cumulation of the value function over time is nothing else but the account. By pricing resources and time, the intelligent PA can do the accounting.

The PA does resource allocation in the micro-economic sense. It can borrow and lend. This can be taken to a sophisticated level, where the PA tries to predict missing information, for example a value function to which it has no access.

The above model of a PA assumes rational behaviour. This is not a defect. The individual PA is optimizing several subjective quantities. Any optimizing strategy is rational. In fact, some economists have *defined* rationality as optimizing behaviour.

6.2.5 An Example

This example traces the functions of three resources during one day. The resources are: my own time, my computer, and my personal library. The other automatic personal assistants involved are my wife (W), the head of department (H), the director of undergraduate studies (U), a PhD student (P), and an employee of a company for whom I consult (B). The access levels are -, r, and w, where - stands for no access, r for read access, and w read-and-write access. An access of rwr-r means read access for my wife, write access for head of department, read access for director of undergraduate studies, no access for PhD student, and read access for the employee. The value $v(t)$ is cumulative over the time slot. It takes into account earnings, but also subjective gains.

These examples have been specifically constructed to illustrate the flexibility of the intelligent PA. Write access to l means that the location can be changed. The head of department can change the location where I have lunch, but not the PhD student (he has read access, and may join me). That my wife can decide what car I take to work, is also indicated by write access to the location. The head of department can decide to put my computer elsewhere: he has access to its location. The value of a resource is under read access for other PAs that need to know it. If someone wants to use my computer, he needs to know the value I attach to it, so that he can pay me for the use he makes. No access to the value means that the resource cannot be used in a negotiation. The director of undergraduate studies has no access to the value of my computer at moments when undergraduate students cannot use the computer. Remark that the Director of undergraduate studies represents the undergraduate students here. In a more realistic situation, all undergraduates would also have PAs, and the access list would be much longer. The employee has write access to the value of my work time, because he pays for it! Read access to the value of my library means that borrowing is allowed. No access means that borrowing is not allowed. When I mention a person here, I actually mean a person's PA.

The access to the description d works in the same way. A person with write access to the description of my time can decide what I do during that time slot. Read access to the computer's description means that the other PA is allowed to

t	l(t)	access	v(t)	access	d(t)	access
8.35	VW	wr—	-20	wrrrr	car trip to college	rrrrr
9.15	9	w—-	-1	wrrrr	deposit L in nursery	wrrrr
9.25	802	rwrrr	1	rrrrr	email and www	rrrrr
9.35	802	rwwww	10	rrrrr	admin: interview rejects	rr—
10.30	508	rrwrr	15	rr-rr	fuzzy systems course	rrrrr
11.20	508	rwwrr	1	rrrrr	available for students	rrrrr
11.30	802	rwwrr	15	rr-rr	prepare next lecture: λ cuts	rrrrr
12.30	SCR	wwwrr	-3	r—-	lunch	wwwrr
13.00	802	rrr-w	25	r—-	consultancy: timetable	r—r
14.00	802	wwrrr	30	rrrrr	research Lagrangian formalism	rrrrr
16.00	805	rrrwr	22	rrr–	meeting A, discuss energy function 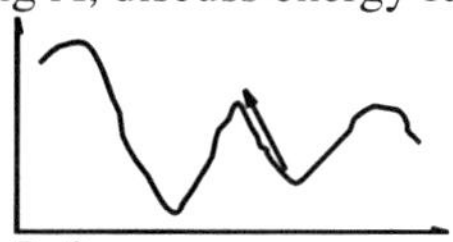	rrrwr
17.30	9	w—-	-2	wrrrr	collect L from nursery	wrrrr
17.45	VW	wr—	-20	wrrrr	car trip home	rrrrr
18.25						

Table 6.1: PDW's diary.

t	l(t)	access	v(t)	access	d(t)	access
0.00	802	—	-1	-r-r-	available for group batch jobs	-r-r-
9.25	802	rwrrr	1	rrrrr	email and www	rrrrr
9.35	802	rwrrr	-5	-r-r-	available for group batch jobs	-r-r-
11.30	802	rwrrr	15	rrrrr	prepare lecture λ cuts slides	rrrrr
12.30	802	—	-1	-r-r-	available for group batch jobs	-r-r-
24.00						

Table 6.2: PDW's computer.

t	l(t)	access	v(t)	access	d(t)	access
0.00	802	rrrrr	0	—	*index of library*	rrrrr
9.35	802	rrrrr	0	—	*index of library*	rrrrr
10.30	508	rrrrr	1	—	*index of library*	rrrrr
13.00	802	rrrrr	1	rrrrr	*index of library*	rwrwr
17.30	802	rrrrr	0	—	*index of library*	rrrrr
24.00						

Table 6.3: PDW's library.

know what processes are running on the computer. Write access to the index of the library means that a book can be borrowed.

Tables 6.1-6.3 describe the resource at a day in the future. During that day, the intermediating agents will turn the description d into a trace of what has happened. The computer will log the commands, processes, and the email. The planned lecture will become the notes of the lecture. The slides to be prepared will become the slides themselves, stored in electronic form. The car trip will become a trace of the trip, the mediating agent being a GPS (global positioning system). The index of the library will become the modified index, after borrowing.

The value will also change after the day has passed. An activity may get a greater value if it has been unexpectedly successful. The value of a lunch may initially be negative (it costs you money), but if you can strike a deal during a business lunch, the value may become positive. A computer that is not being used has a negative value, it costs you money. But if several users make use of it, the value can become positive. In summary, the value has a dual purpose. It can be used as a negotiating tool, for example when an activity has to be rescheduled and replaced by another one. Its other purpose is to do accounting. At the end of the day, the sum of the values (or, the integral of $v(t)$ over t) indicates a gain or a loss.

In case the timetables seem difficult to specify and too detailed, it has to be kept in mind that they are used by an automatic PA only. The actual table would be constructed via a user-friendly programme (another mediating agent), that could account automatically for the many recurrent activities as lectures, trips to work, etc.

6.3 Finding a Stable Strategy

6.3.1 The Discrete Event Simulator

The program simulates a number of personal assistants that interact with one another. For every moment in time, the PA has a location, a cost of its time, a description of its state, and an accumulated value ("money"). The time is discrete in one-minute intervals.

The PAs interact by attempting to change each other's locations and descriptions. In employee-terms, changing a *location* means that one employee tells another to go to a particular location during a period in the future. Changing the *description* means telling somebody what to do. These changes are not always allowed: there are access restrictions, e.g. only certain employees can tell you what to do. A PA who changes another PAs location or description has to pay for this, according to the cost of the other PAs time. So the PAs earn money from other PAs that are telling them what to do, and they spend money by telling other PAs what to do.

The period, during which a PA wants to change another PA's location, descriptor, or both, is uniformly distributed between 2 and 60 minutes. The change will start at a certain moment in the future. This moment too is uniformly distributed between 0 (immediate change) and a maximum value.

There are as many locations as there are PAs. Some PAs can be in the same location, e.g. during a meeting. Every PA has write access to its own location, value, and descriptor (l, v, and d). The descriptor contains at most 10 words per minute. The number of words is uniformly random and the upper limit is inversely proportional to the length of the period (in minutes) during which a PA is committed to a location. If a PA changes another PAs descriptor, the number of words added per minute is again random and the upper limit is inversely proportional to the period for which the changes are made.

The access from a PA to another PA's location is uniformly distributed between no access, read access, or write access. The access to another PA's time value is uniformly random too, but it has to be at least as strict as the access to the location, and it is never write access. The access to the descriptor is random, but it is at most as strict as the access to the location. So the access restriction to the descriptor is the weakest, and the access restriction to the time value is the strongest.

Every minute, one PA tries to change the l or d of another PA. Which PA is chosen, depends on a *strategy*.

6.3.2 A Stable Strategy

The access restrictions determine which PAs can interact which each other, but they also leave a degree of freedom to the PAs. The PA can exploit this to maximize some personal, local goal. This means that the PAs act as economic agents.

We will use the accumulated value

$$m_i(t) = \sum_{t'=0}^{t} v_i(t') \tag{6.1}$$

to determine whether PA i can act. The accumulated value m_i has to be positive, otherwise PA i cannot change anything to another PA.

If i changes $l_j(t)$ from t_1 to t_2, then $m_i(t)$ decreases with $v_j(t)$ every minute (PA i "pays"), and $m_j(t)$ increases with $v_j(t)$ (PA j "receives"). This means

$$m_i(t_2) - m_i(t_1) = -\sum_{t'=t_1}^{t_2} v_j(t'), \tag{6.2}$$

and

$$m_j(t_2) - m_j(t_1) = \sum_{t'=t_1}^{t_2} v_j(t'). \tag{6.3}$$

If the descriptor is changed instead of the location, the change per minute to m_i is $-v_j(t)$ multiplied by the fraction of the total descriptor length of d_j that i wants

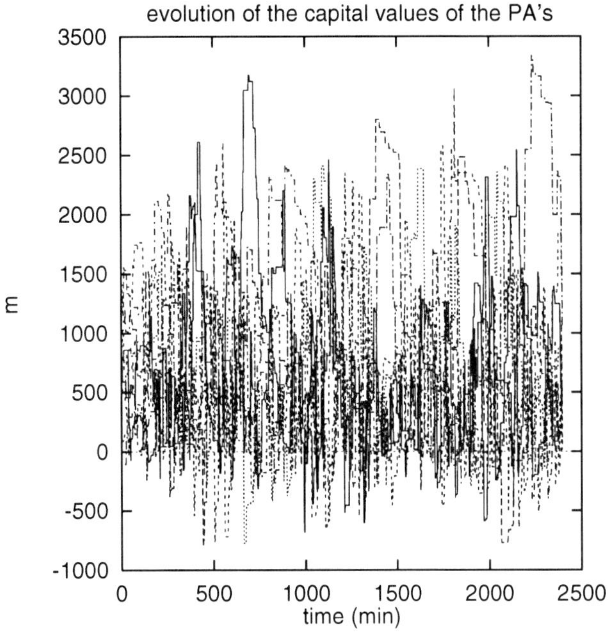

Fig. 6.1: The accumulated values m_i for a network of 10 PAs, simulated over a week. The PAs continue earning and paying, the system does not get stuck.

to add. The change to m_j is similar, but has opposite sign. If both location and descriptor are changed, the changes in m_i are as for the location only.

The balancing of accumulated value in this strategy (paying and receiving) implies that the sum of the accumulated values of all PAs is constant:

$$\sum_i m_i(t) = c. \tag{6.4}$$

This is a conservation law. The initial values $m_i(0)$ are uniformly distributed between 0 and the maximum time it can take between the initiating of a change and the end of the actual change. This is an arbitrary choice, and it is part of our strategy.

Initially, the values are uniformly distributed between 1 and 10. The strategy is as follows. *The PA i that initiates a change looks for the cheapest PA j that can do the change over the given period, given the access restrictions. PA j increases its m_j, and adds 1 to its values for all moments in time. If the initiating PA has now $m_i < 0$, it halves all the values of its time.* This can be expressed as follows, in a simplified way. The cheapest PA for a job is always chosen. The one that got the job increases its rates by 10% after the job. If a PA goes bankrupt, it halves its rates. This strategy we will call free market with 10% inflation and 50% discounting, abbreviated $FM(10,50)$.

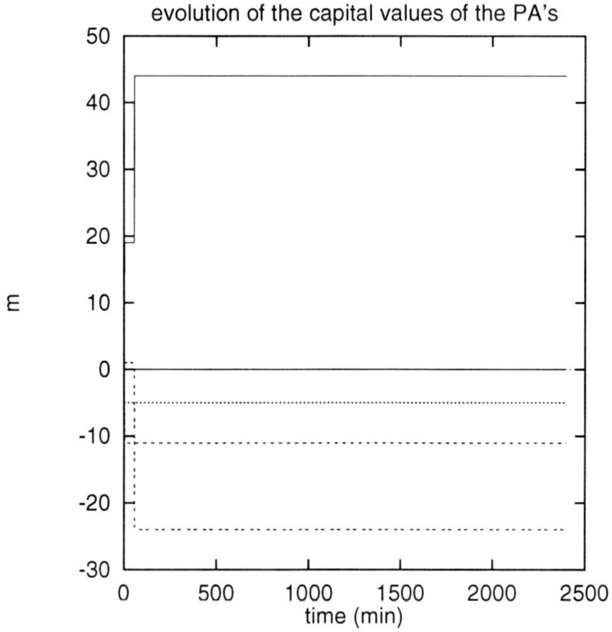

Fig. 6.2: A network of 10 PAs, simulated over a week. The cheapest PA is not chosen, and the initial endowment is 1000 times smaller than in Figure 6.1. The m_i do not change any more; the PAs are stuck in a trivial equilibrium where their location does not change any more, and they can only change their descriptor themselves.

We can now look at the evolution of the accumulated values, or capital values, of the PAs. Figure 6.1 shows a simulation for 10 PAs, during one week (5 days of 8 hours). Changes can take effect up to 3 days in the future. The changes in the m_i reflect changes in the PA's location and descriptor, the PAs are "moving around" and "doing things". This we will call a *nontrivial equilibrium*. We have commented on the nature of the equilibrium in Section 5.1. The nontrivial equilibrium means that not all agents are "frozen", some agents (not necessarily the same ones over time) keep trading. Calculating the equilibrium distribution π_j from Theorem (3.6) is not feasible, because there are too many different states. However, the PAs trade without taking the past states into account, so we are assured of the Markov property. This assures us of the equilibrium, and Figure 6.1 assures us of the nontrivial equilibrium, the agents keep trading. We will see later that this is independent of the time limit of the simulations.

We will show via simulations that the strategy $FM(10,50)$ has to be changed significantly for this nontrivial equilibrium to disappear. In other words, we will show that the strategy $FM(10,50)$ is stable. Stability here is meant in the sense of an equilibrium distribution of a stochastic process. We will make this clear later, in the mathematical model.

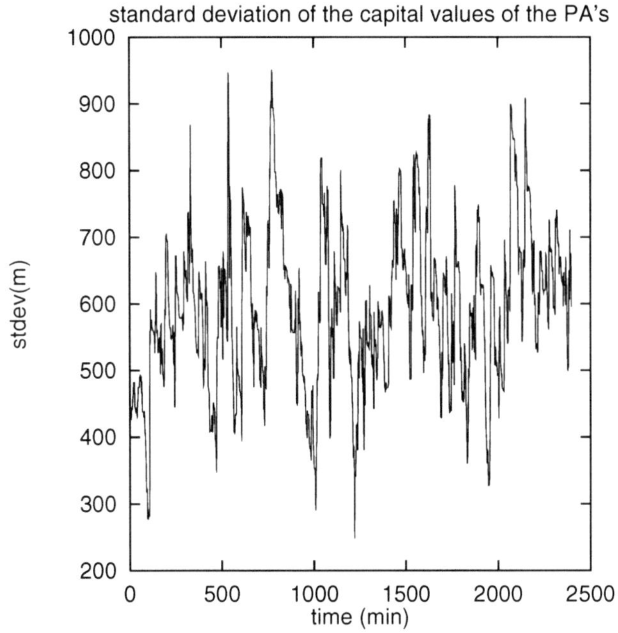

Fig. 6.3: The same network as in Figure 6.1, but now showing the evolution in time of the standard deviation of m_i.

The trivial equilibrium occurs when the PAs run out of money ($m_i < 0$), and they are all frozen in their locations, and they can only change their own descriptors. In an organization this would mean that the employee's do not interact any more, they just sit at their desks and everybody decides for himself what he is going to do. In this case the values m_i stay constant. This is illustrated in Figure 6.2.

The conditions are the same as in Figure 6.1, but the strategy $FM(10,50)$ had to be altered significantly. The cheapest PA is not chosen any more, just the first PA whose access restrictions allow the changes. The initial endowments $m_i(0)$ are 1000 times smaller than in $FM(10,50)$. Just to show how stable $FM(10,50)$ really is, initial endowments 100 times smaller would still give a nontrivial equilibrium as in Figure 6.1.

It is important to check the behaviour of a strategy such as $FM(10,50)$ over a longer period, to see whether all PAs do not accidentally run out of money at the same time. In that case, a trivial equilibrium would occur. In order to have a clearer graph, we will plot the standard distribution of the $m_i(t)$ at a certain moment t in time. Figure 6.3 shows this for 10 PAs, all other parameters and the strategy are the same as for Figure 6.1.

If the simulated period is three months, with changes taking effect at last within a week (you can book somebody's time at most a week beforehand), we obtain Figure 6.4 for the standard deviation of the m_i. The equilibrium continues to be nontrivial.

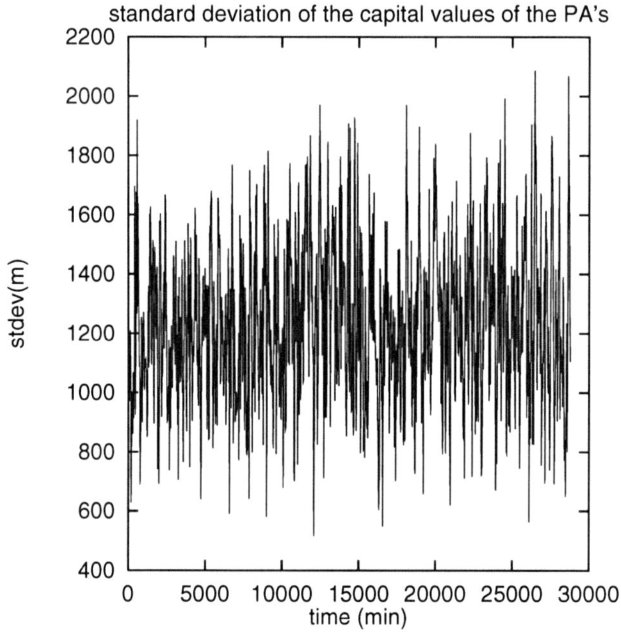

Fig. 6.4: A network of 10 PAs, simulated over three months, with the free market strategy $FM(10,50)$. The equilibrium is nontrivial, the m_i fluctuate, indicating that the locations and descriptors change.

This indicates that it is a good choice to have the initial endowment uniformly distributed between 0 and the maximum time it can take between the initiating of a change and the end of the actual change.

If the initial endowment is smaller, the strategy is still stable, provided that the cheapest PA is chosen. This is illustrated in Figure 6.5 for an initial endowment 1000 times smaller than in Figure 6.1.

The equilibrium is nontrivial. If any available PA is chosen, we obtain the trivial equilibrium of Figure 6.2.

Figure 6.3 shows that the difference between "rich" and "poor" PAs stays approximately the same during the simulation. This becomes different when the bankrupt PAs do no discounting, a strategy $FM(10,0)$. Figure 6.6 shows that the standard deviation of the capital or accumulated value increases. Some PAs will occasionally be severely bankrupt (remember that the sum of m_i over all PAs stays constant), but they will be able to overcome this, because the other PAs are choosing the cheapest PA they can find. The equilibrium is nontrivial, even without discounting, only the difference between "rich" and "poor" PAs increases.

Simulations over one week are sufficient to indicate this trend. The simulations over three months in Figure 6.7 confirm this.

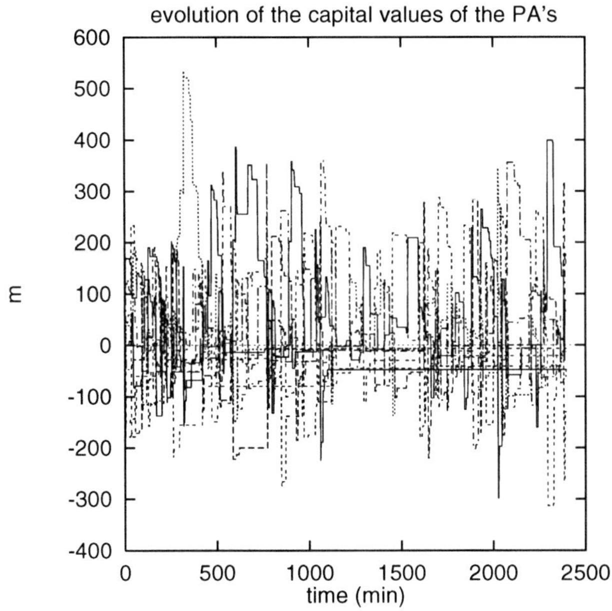

Fig. 6.5: A network of 10 PAs, with initial endowment 1000 times smaller than in Fig. 6.1, but with the cheapest PA chosen.

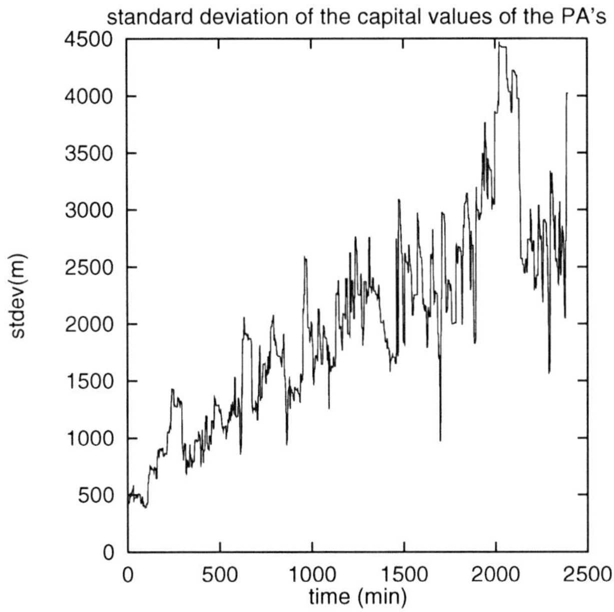

Fig. 6.6: The network without discounting on bankruptcy. The standard deviation of the m_i increases.

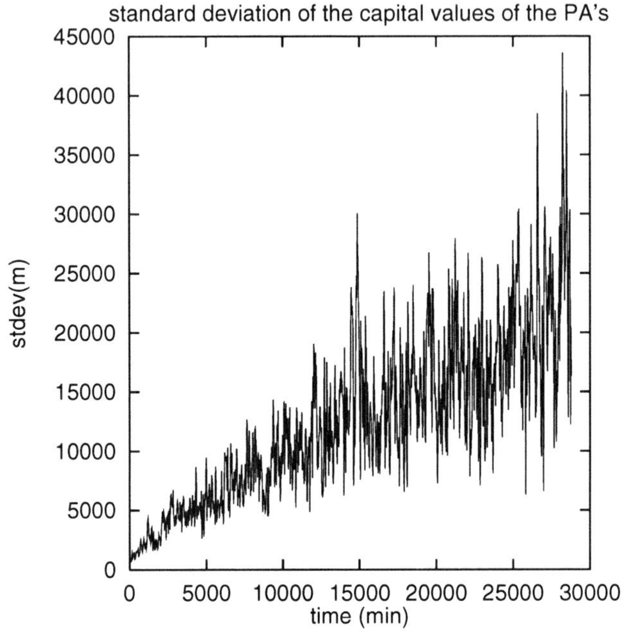

Fig. 6.7: The same network as in Figure 6.6, but now simulated over three months instead of one week. The increase of the standard deviation is confirmed.

6.4 Conclusion

We have presented a realistic model of an intelligent personal assistant. A network of such assistants will continue operating without deadlock if the personal assistants can buy and sell time slots in a free market with inflation and discount on bankruptcy. We found that this strategy is robust. This result is complementary to [Kephart et al., 1997].

If time slots are seen as a paradigm for general resources, we have found a strategy to coordinate exchange of resources in a distributed system. This strategy was based on a micro-economic model of the distributed system. Do our PAs trade in an efficient market? Yes, if we adopt the definition of market efficiency of Section 4.5.2. As the PAs keep trading, the incoming requests for time slots will be fulfilled to some extent. We cannot show that our trading strategy for the PAs is optimal. What we have been able to show is that the equilibrium is nontrivial, i.e. the PAs keep trading. Note that the definition of market efficiency from 4.5.2 is related to the definition of efficient production, as it is used in economy [Mas-Colell et al., 2004]. In 5.1 we have mentioned the notion of efficient trading in the stock market [Elton et al., 2006]. We cannot show that our market of trading PAs is efficient in terms of an efficient stock market, because we have restricted our trading algorithms to those that ignore the past. Investigating all possible trading algorithms, also those that are non-Markov , would require considerable more work. We do not consider this worthwhile, because we have found a robust algorithm that uses only the current state of the system: a free market with inflation and discount on bankruptcy.

The number of agents was assumed constant. A varying number of agents requires a different treatment, using evolutionary dynamics, see Section 3.3.2.1 and [De Wilde, 2002; De Wilde et al., 2003].

Chapter 7
Conclusions and Future Work

Complex systems such as societies, markets and biological systems can be effectively modelled using multi-agent systems. Motivated by the need to understand the dynamics as well as analyse the properties which emerge from the interactions that occur in such systems, this work contributes to the fields of convergence and knowledge exchange in multi-agent systems. This chapter reviews the book's scientific contributions (Section 7.1) and then discusses promising directions for future research in this challenging class of domain (Section 7.2).

7.1 Contributions

The four main contributions of this book are summarised as follows:

1. A definition of stability in multi-agent systems has been proposed after the fundamental concepts behind the notion were investigated. The system is viewed as a discrete time Markov chain with a potentially unknown transition probability distribution. It is considered to be stable when its state has converged to an equilibrium distribution. The definition proposed is the only one which takes into account the game nature of multi-agent systems, is relevant to systems with a varying number of agents and is supported by the mathematical framework of stochastic systems. Several artificial multi-agent ecosystems were developed and their properties were analysed in order to confirm the validity of the proposed definition.
2. It is shown that in an ecosystem of networked businesses: (a) A straightforward set of assumptions is enough to give rise to information exchanges among the companies, and (b) This information exchange is shown to increase the efficiency of the market. The importance of these findings lies in the fact that even though high-level concepts such as network effects, reputation and trust are not taken into account in the model of the market used for the simulation, information exchange naturally emerges as a successful profitable behaviour. Knowledge exchange is known to be beneficial for industry, but in order to explain it, authors

M. Chli and P. De Wilde, *Convergence and Knowledge Processing in Multi-Agent Systems*, Advanced Information and Knowledge Processing,

have used high-level concepts. The investigation we performed in a setting of software providers offers a plausible and elegant explanation of how and why organisations adopt information exchange and why it benefits the market as a whole when this happens.

3. A method of optimising the performance of a system with information provided through interaction with it is proposed. The results of this work are illustrated in the context of the Internet retrieval process. The expansion of the Internet has made the task of searching a crucial one. Internet users, however, have to make a great effort in order to formulate a search query that returns the desired results. Many methods have been devised to assist in this task by helping the users modify their query to give better results. In this work we propose an interactive method for query expansion. It is based on two observations: (a) documents are often found to contain terms with high information content which can summarise their subject matter and (b) some users are more proficient in formulating accurate queries than others. We present experimental results which demonstrate that our approach significantly shortens the time required in order to accomplish a certain task by performing web searches. Furthermore, knowledge exchange between individual searchers is utilised to make for more efficient search sessions.
4. By using a multi-agent based approach, and after developing the notions of equilibrium and convergence in Chapter 3, we were able to find a strategy for the agents in an important resource management and scheduling problem. By assigning values to resources, we proposed a micro-economic trading strategy for the PA's. This is a free market strategy with inflation and discount on bankruptcy. For an observer of the current world economy, it is probably no surprise that we found a free market strategy to be convergent and robust. In Chapter 4 we found inspiration in the Marseille fish market, in Chapter 6 we found inspiration in the global world economy. We do not claim to have improved economies. Instead, we have taken inspiration in economics to improve the mechanisms of multi-agent systems, and the strategies of individual agents. The applications we have developed throughout this book: trading, load assignment, software services, and scheduling are wide ranging. With small adaptations the reader should be able to devise convergent and robust strategies for his or her problem, too. The authors keep working in this exiting area, and are always willing to give advice.

We have shown in the book that Markov models are sufficiently complex to model a wide range of applications, and to analyse the convergence. Of course our simulations have not shown all possible behaviours that can occur. Now that the link between Markov process and multi-agent systems has been established, however, the whole realm of results on Markov chains, developed over the last century, can be used. If the reader analyses a multi-agent system using our techniques, and observes behaviour that is not covered in this book, then he or she should consult a book on Markov chains. Cox' book [Cox and Miller, 1984] is an old favourite, and still very useful. We have used it extensively. Ross [1996] is more modern, and very useful for the practitioner as well as the student. Stirzaker [2005] can also be recommended. Physicists have developed a good understanding of pathological behaviour in stochastic systems, and Gardiner [2004] will be useful.

7.2 Future Directions

Much of our current work has been focused on empirical understanding of complex systems. We have investigated, using agent-based modelling, the reasons that cause complex systems such as markets, societies and ecosystems to exhibit certain macro-regularities. In the future, we would like to extend our work to achieve normative understanding of such systems. In other words, we want to create realistic, accurate models that will enable us to make decisions relevant to the real-world systems we investigate, in order to achieve optimal results. These decisions may concern design choices (e.g. market mechanisms, economic policies, infrastructure decisions) or parameter values (e.g. level of taxation in a market).

A realistic model will give us qualitative insight of the complex system under investigation and it will enable us to generate theories that can be used in a predictive manner. This will help clarify existing macroscopic phenomena and possibly predict new ones. Finally, in this process we will also be making contributions to the methodology of agent-based modelling.

7.2.1 Ecosystems of Networked Businesses

Networked business ecosystems are believed to be the future of business to business (B2B) relationships. The formation of Internet business communities will enable companies to collaborate through loosely coupled business services. Participants will register business services that others can discover and incorporate into their own business processes. Companies will be able to build on each other's services, creating new services and linking them into business models that will potentially transform the industry [Tenenbaum and Khare, 2005].

- Further work needs to be done to address critical issues that concern the infrastructure of these networks, inter-enterprise integration and collaboration, in order for business ecosystems to materialise. Multi-agent systems provide an ideal platform for modelling networked business communities and evaluating different architectures and hierarchies [Nahm and Ishikawa, 2005].
- The investigation of the interactions among the companies in a networked business ecosystem constitutes another avenue for future work. Inspiration can be drawn from ecology and biological processes where an entity or a structure is progressively modified to give better performance in an environment as well as from social network analysis, which deals with interactions inside networks of humans.
- Issues of cooperation and competition among companies in a Network Business Ecosystem are of particular interest. The term co-opetition, defined as simultaneous cooperation and competition was introduced by Brandenburger and Nalebuff [1996]. Co-opetition was correlated to strategic alliances between European SMEs in a case study produced by Eikebrokk and Olsen [2005]. In a network

business ecosystem, cooperation and competition will be significantly facilitated. Using multi-agent systems as well as ecology theories concerning symbiosis and competition the effect on businesses as well as the market can be investigated.
- Finally, placing this work in an actual business context e.g. travel industry, supply chain management, load distribution and transportation will help us draw conclusions that can be used in a predictive way.

7.2.2 Exchange in Natural Ecosystems

Biological reserves created to conserve endangered species as well as issues concerning their size in area and the number of species they contain are the subject of recent research. The inhabitants of these nature reserves are evolving populations rather than static biological entities. Specialists are interested in determining the factors that affect sustainment costs along with attaining both the ecological and economic optimum when establishing a new nature reserve. We intend to apply our findings regarding exchange and adaptation of populations described in Chli and De Wilde [2009] and De Wilde et al. [2003] to investigate this problem.

Economic considerations in the optimal size and number of reserve sites were studied by Groeneveld [2005] . The debate among ecologists on the optimal number of reserve sites under a fixed maximum total reserve area is known as the "single large or several small" (SLOSS) problem. Along similar lines Hamaide et al. [2005] studied the trade-off between species abundance and species diversity in biological reserves. Requiring representation of each species in at least one parcel in the system and seeking the minimum number of parcels in the reserve system to achieve this is termed the 'species set covering problem' (SSCP). Factors that are taken into consideration are the ability to hold as many species as possible and have low extinction rates while keeping the costs of nature conservation low.

In De Wilde et al. [2003] it was shown using population dynamics that if maintaining diversity is to be successful, i.e. without lowering too much the payoff for the non-endangered strategies/species, it has to go on forever, because the non-endangered strategies still get a good payoff, so that they continue to thrive.

In Chli and De Wilde [2009] population exchange was shown to increase the efficiency of a market in the sense that all the needs were equally satisfied. We intend to combine our results of Chli and De Wilde [2009] and De Wilde et al. [2003] to examine what effect they have in the context of diversity in a population as well as investigate how they can be applied in order to provide answers to problems such as SLOSS and SSCP.

7.3 Concluding Remarks

Empirical results validate all four contributions within a number of domains. The generality of the contributions is verified by applying them to simulations of complex market, social and biological systems. Ultimately, this work demonstrates the fact that interactivity/exchange and convergence in multi-agent systems are issues that are significantly interrelated with direct influence to each other.

Appendix A
The EEII Project

The EEII project focused on exploring the possibility of using ecological methods within the development of agent-based information systems. The project web page can be found at http://www.ee.ic.ac.uk/research/neural/eeii.html. The partners were Imperial College London, Eindhoven University of Technology and Universidade Nova de Lisboa. Within the project, agent systems were studied from an ecological point of view and properties of agent systems such as scalability [Marwala et al., 2001], openness [Abramov et al., 2001], adaptability [Simoes-Marques et al., 2003] and stability [Chli et al., 2003][1] were investigated.

Scalability [Marwala et al., 2001] is viewed as the relationship of any parameter with respect to the number of agents in the population. For the purpose of this investigation a market environment was simulated with the traders modelled as agents. Soft computing (neural networks and fuzzy algorithms) as well as genetic algorithms were utilised to model the decision making component of the traders. It was shown using simulation that the number of agents participating in the trading game modelled is directly proportional to the computational time taken to run the simulation. It was further observed that no player had a monopolistic advantage on the prediction of the market.

The concept of openness [Abramov et al., 2001] of a multi-agent system allows for more than one interpretations. One of the contributions of the EEII project has been a classification of the different types of openness as well as a generic definition. An agent system has being developed as a basis for the study of openness-related properties. This agent system was used to conduct a set of experiments to allow the study of several types of openness.

Adaptability is the ability of the agents to adapt in a dynamic environment. In Simoes-Marques et al. [2003] an agent model of a market buyer/seller scenario was developed. The agents adapt in order to maximise their fitness using both classical and fuzzy computing methods. Sellers seek to maximise the profit they obtain on trades, while buyers seek to maximise the degree of acceptance of the goods they

[1] Our contribution and the part of the EEII project included in this book is the investigation of stability of multi-agent systems.

buy. This degree is affected by price and user preference constraints. Both classical and fuzzy computing methods are used for the adaptation of agents. The proposed model was shown to be effective as adaptable agents were more successful in the competitive trading environment.

Artificial adaptation of populations was studied in De Wilde et al. [2003] with simulations of an ecosystem of interacting software agents. The agents have a strategy, and receive a payoff for executing that strategy. Unsuccessful agents become extinct. The repercussions of maintaining a diversity of agents were investigated.

It was shown that if maintaining diversity is to be successful, i.e. without lowering too much the payoff for the nonendangered strategies, it has to go on forever, because the nonendangered strategies still get a good payoff, so that they continue to thrive, and continue to threaten the endangered strategies. For example in the animal and plant kingdoms, the number of endangered species seems much smaller that the number of nonendangered ones, although there is great uncertainty on the numerical data. In this situation, it is shown using replicator dynamics that it is possible to conserve the endangered species, if the effort is spread over all other species.

On the other hand, it is shown that this is not sustainable if the number of endangered species is of the same order as the number of nonendangered ones. In other words, one should not try to control the pure Darwinian evolution in a population of competing agents by artificially maintaining a diversity of agents. If the number of endangered species is much smaller than the others, they will have little influence on the dynamics of the system, and whether the others sustain them or not will make little difference again.

Appendix B
Statistical Analysis

Some concepts from statistical hypothesis testing are explained here. An effort was made to make this chapter self-contained but the reader is encouraged to refer to any statistics textbook such as Fisher [1970]; Crawshaw and Chambers [2001] for more in-depth information.

B.1 Statistical Hypothesis Testing

Statistical hypothesis is the assumption that a parameter of some distribution has a certain value. *Null hypothesis* H_0 is a hypothesis we may test by accepting it in the first place. *Alternative hypothesis* H_1 is the hypothesis, which is accepted once the null hypothesis is rejected. It is therefore the assumption that H_0 is wrong. The object of a statistical test is to test the value of a population parameter using a random sample and observing the corresponding sample parameter. To do so, a function of the sample parameter of a known probability distribution may be used, called the *test statistic*. By accepting the null hypothesis we establish that the test statistic has a certain probability to lie within a certain range considered likely. The probability to lie outside this certain range, considered unlikely, is known as the *significance level*. If, given that H_0 is true, the observed value of the test statistic is in a range that it has small probability to lie (equal to the level of significance, e.g. 0.05 or 5%) then we reject H_0 and conclude that the test is significant. Otherwise we accept H_0 and resolve that the test is not significant.

B.2 Tests for Showing That Two Samples Come from the Same Distribution

To determine whether the two samples come from the same distribution a *t-test* for the mean and then an *F-test* for the variance are performed. The methods used are described below.

The tests are performed on two samples of the "Tasks executed per trader" metric from the unstable system experiment described in Section 3.4.1.3. The data obtained for each of the metrics in this experiment is split in two groups. Sample 1 contains data for rounds 1999 to 5999 and Sample 2 contains data for rounds 6000 to 10 000. Any data recorded at the start of the experiment, which is regarded as a burn-in period, is not taken into account. The methods for carrying out the tests for the mean and variance are illustrated below.

The above tests are mainly intended for normal populations. However, they may be used for non-normal ones provided that the sample size is sufficiently large. This is because according to the central limit theorem, even if the population is non-normal, $\overline{X}$ is appropriately $\mathcal{N}(\mu, \frac{\sigma^2}{n})$. Furthermore, if the population variance σ^2 is not known it may be replaced by $\hat{\sigma}^2 = \frac{\sum x^2 - n\bar{x}^2}{n-1}$ for large samples. Our samples are of the order of 1000, therefore they can easily be considered as large samples.

1. **Test for the variance**
   H_0: Two unpaired random samples (Sample 1 and Sample 2) come from populations of equal means.
   H_1: Sample 1 and Sample 2 come from populations of unequal means

   If two random samples (not necessarily of the same size) are $\{X_1, X_2, \ldots, X_n\}$ and $\{Y_1, Y_2, \ldots, Y_m\}$ the best unbiased estimators for the population variance, in each case, are

   $$S_x^2 = \frac{\sum(x-\bar{x})^2}{n-1} \quad \textit{and} \quad S_y^2 = \frac{\sum(y-\bar{y})^2}{m-1}$$

   Then the test statistic $F = \frac{S_x^2}{S_y^2}$ has a $F(n-1, m-1)$ distribution, i.e. an F-distribution with degrees of freedom $n-1$ and $m-1$, the first parameter corresponding to the numerator and the second to the denominator.
   H_0: The variance for task/traders samples 1 and 2 is the same.
   H_1: The variance is different.

   We calculate:
   $S_x^2 = 2.04 \times 10^{-3}$ and $S_y^2 = 1.86 \times 10^{-3}$, so we obtain $F = \frac{S_x^2}{S_y^2} = 1.095$. The critical values for $F_{0.05}(4000, 4000)$ are 0.94 and 1.06. F-distribution is associated with the χ^2-distribution which, contrary to t-distribution, is not symmetric. As a result we need to consider separately the upper and lower critical values for a two-tailed test.

The test statistic lies in the critical region so we may say that the test is significant at the 5% significance level and that the two samples are not drawn from populations with the same variance.

2. **Test for the means**
Since the variances of the two samples are not known, the test statistic will be

$$T = \frac{\overline{X} - \overline{Y}}{S_p\sqrt{\frac{1}{n} + \frac{1}{m}}} \quad where \; S_p = \frac{(n-1)S_x^2 + (m-1)S_y^2}{m+n-2}$$

The variable T is not a normal variable but has a t-distribution with parameter $\nu = n - 1$, i.e. $T(n-1)$. It is worth noting here that as $\nu \to \infty$ the t-distribution approaches the normal distribution.
We calculate:
$\overline{X} = 0.067, \overline{Y} = 0.066, S_x^2 = 0.002, S_y^2 = 0.0019$ and therefore $S_p^2 = 0.002$ and $T = 0.553$.
In this case the question is focused on whether the means are different. Hence the critical region (shaded) in Figure B.1 has to be two sided. Testing at the 5% significance level yields two critical values ± 1.96 and T does not lie in the critical region so we may say that there is no significant evidence that the populations' means are different.

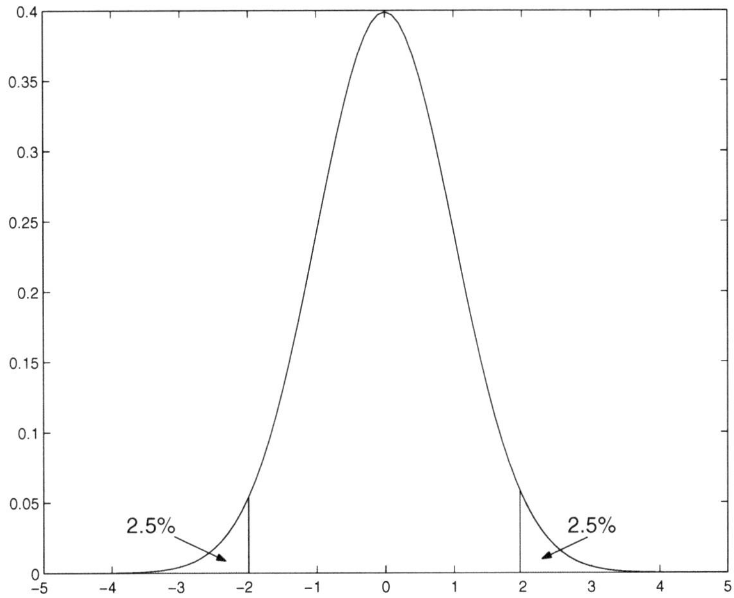

Fig. B.1: Z distribution for 8000 degrees of freedom. The critical region for the 5% significance level.

The test for the means succeeded. However, the test for the variances did not succeed. This means that there is no significant evidence that the two samples come from the same distribution. Therefore, the system is *not stable* when run with the initial conditions mentioned above.

Similar tests were performed for all of the metrics, which show analogous results. In general, in order to accept that a system is stable it is required that all of the metrics exhibit stability.

Appendix C
Methodology: Evolutionary Algorithms

In order to model evolution in populations as well as learning we have used several evolutionary algorithms in our model. In this section we give a brief overview of these algorithms.

Evolutionary algorithms [Brazdil, 1993] "is an umbrella term employed to describe computer-based problem solving systems which use computational models of some of the known mechanisms of evolution as key elements in their design and implementation." A variety of evolutionary algorithms have been proposed by several researchers. The major ones are: genetic algorithms, evolutionary programming, evolution strategies, classifier systems and genetic programming. They all share a common concept of simulating the *evolution* of objects/structures using the processes of selection, mutation and reproduction. The processes depend on the performance/fitness of the individuals under consideration as defined by their environment and quantified by a fitness function.

According to Spears [2000], "evolutionary algorithms maintain a population of structures that evolve according to rules of selection and other operators, that are referred to as search operators (or genetic operators), such as recombination and mutation. Each individual in the population receives a measure of its fitness in the environment. Reproduction focuses attention on high fitness individuals, thus exploiting the available fitness information. Recombination and mutation perturb those individuals, providing general heuristics for exploration. Although simplistic, these algorithms are sufficiently complex to provide robust and powerful adaptive search mechanisms".

A **genetic algorithm** (GA) [Goldberg, 1989] is a model of machine learning inspired by the mechanisms of genetics, which has been applied to optimisation. It operates with an initial population containing a number of trial solutions. Each member of the population is evaluated (to yield a fitness) and a new generation is created from the better of them. The process is continued through a number of generations with the aim that the population should evolve to contain an acceptable solution. In Ribeiro Filho et al. [1994] it is stated that GAs are particularly suitable for solving complex optimization problems and hence for applications that require adaptive problem-solving strategies. In order to make genetic algorithms reach an optimal solution faster, parallel implementations of GAs are often used [Cantu-Paz, 1998]. Genetic algorithms are used for a number of different application areas. According to Sumathi et al. [2008], "an example of this would be multidimensional optimisation problems in which the character string of the chromosome can be used to encode the values for the different parameters being optimised".

In practice, therefore, we can implement this genetic model of computation by having arrays of bits or characters to represent the chromosomes [Sumathi et al., 2008]. Simple bit manipulation operations allow the implementation of crossover, mutation as well as other operations. Crossover involves combining strings to swap values, e.g. $101001 + 111111 \rightarrow 101111$. Mutation involves spontaneous alteration of characters in a string, e.g. $000101 \rightarrow 100101$. Although a substantial amount of research has been performed on variable-length strings and other structures, the majority of work with genetic algorithms is focused on fixed-length character strings.

Statistical classification is a type of supervised learning algorithm which takes a feature representation of an object or concept and maps it to a classification label. A classification algorithm [Winkler et al., 2007] is designed to learn, or in other words, to approximate the behaviour of a function which maps a vector of features $[X_1, X_2, ..., X_n]$ into one of several classes by looking at several input–output examples of the function.

An instance of a classification algorithm is called a classifier. Learning classifier systems [Holland, 1976] are a machine learning technique which combines evolutionary computing and reinforcement learning to produce adaptive systems. It is a minimal form of modelling learning in the sense that it is not necessary to make assumptions about the way the agents perform their reasoning. In addition to that, the absence of any assumptions or biases in the learning process leads to results that can be generalised. A classifier consists of a set of rules, which have a condition C (`if` part) an action A (`then` part) and a strength measure s. An example of a classifier system is shown in table C.1.

`if` C_1	`then` A_1,	s_1
`if` C_2	`then` A_2,	s_2
`if` C_3	`then` A_3,	s_3
`if` ...	`then` ...	...
⋮	⋮	⋮

Table C.1: An example of a classifier system.

In the model described in detail in Section 4.4.2, genetic algorithms and classification algorithms have been used to model evolution of populations of solutions and learning.

References

V. Abramov, N. Szirbik, J. Goossenaerts, T. Marwala, P. De Wilde, L. Correia, P. Mariano, and R. Ribeiro. Ontological basis for open distributed multi-agent system. In *Proceedings of Symposium on Adaptive Agents and Multi-agent Systems*, 2001.

V. Abramov, N. Sziribik, and J. Goossenaerts. Workpackage 4: Openness (IST-1999-10304). Technical report, Technical University of Eindhoven, 2003.

R. Ashri, S. D. Ramchurn, J. Sabater, M. Luck, and N. R. Jennings. Trust evaluation through relationship analysis. In *Proceedings of the Fourth International Joint Conference on Autonomous Agents and Multi-agent Systems*, pages 1005–1011, 2005.

K. J. Åström and B. Wittenmark. *Adaptive Control*. Prentice-Hall, Reading, Massachussetts, 1995.

H. Balakrishnan, M. Stemm, S. Seshan, and R. H. Katz. Analysing stability in wide-area network performance. *ACM SIGMETRICS Performance Evaluation Review*, 25(1):2–12, 1997.

Y. Bar-Yam. A mathematical theory of strong emergence using multiscale variety. *Complexity*, 9(6):15–24, 2004. ISSN 1076-2787. doi: http://dx.doi.org/10.1002/cplx.20029.

J. Bessen. *The Economics of Open Source Software Development*, chapter Open source software: free provision of complex public goods. Elsevier, 2006.

Z. Bodie, A. Kane, and A. Marcus. *Investments, 5th Edition*, chapter 12: Market Efficiency. McGraw-Hill and Irwin, 2002.

R. C. Bodner and F. Song. Knowledge-based approaches to query expansion in information retrieval. In *Canadian Conference on AI*, pages 146–158, 1996.

E. Bonabeau, M. Dorigo, and G. Theraulaz. *Swarm Intelligence: from Natural to Artificial Systems*. Oxford University Press, Inc., New York, NY, USA, 1999. ISBN 0-19-513159-2.

A. Bonaccorsi and C. Rossi. Altruistic individuals, selfish firms? The structure of motivation in open source software. *First Monday*, 9(1), 2004.

A. Bonaccorsi and C. Rossi. Why open source software can succeed. *Research Policy*, 32(7):1243–1258, 2003.

F. Bousquet and C. Le Page. Multi-agent simulations and ecosystem management: a review. *Ecological Modelling*, 176(3-4):313–332, 2004.

A. M. Brandenburger and B. J. Nalebuff. *Co-opetition*. Currency-Doubleday, New York, 1996.

P. B. (ed.) Brazdil. *Editorial, Machine Learning: Proceedings of the European Conference on Machine Learning*. Springer, New York, NY, USA, 1993.

J. Bredin, R. T. Maheswaran, C. Imer, T. Basar, D. Kotz, and D. Rus. A game-theoretic formulation of multi-agent resource allocation. In *Proceedings of the Fourth International Conference on Autonomous Agents*, pages 349–356, 2000.

T. Brenner. *Local Industrial Clusters, Existence, Emergence and Evolution*. Studies in Global Competition. Routledge, London, 2004.

E. Cantu-Paz. A survey of parallel genetic algorithms. *Calculateurs Paralleles, Reseaux et Systems Repartis*, 10(2):141–147, 1998.

K. M. Carley, D.B. Fridsma, E. Casman, A. Yahja, N. Altman, L.-C. Chen, B. Kaminsky, and D. Nave. Biowar: Scalable agent-based model of bioattacks. *IEEE Transactions on Systems, Man and Cybernetics, Part A*, 36(2):252–265, 2006.

M. Chli and P. De Wilde. Workpackage 6: Stability (IST-1999-10304). Technical report, Imperial College London, 2003.

M. Chli and P. De Wilde. The emergence of knowledge exchange: an agent-based model of a software market. *IEEE Transactions on Systems, Man and Cybernetics, Part A*, 2009. Accepted for publication.

M. Chli and P. De Wilde. Internet search: subdivision-based interactive query expansion and the soft semantic web. *Applied Soft Computing*, 2005. Accepted for publication in special issue.

M. Chli, P. De Wilde, J. Goossenaerts, V. Abramov, N. Szirbik, L. Correia, P. Mariano, and R. Ribeiro. Stability of multi-agent systems. In *Proceedings of the 2003 IEEE International Conference on Systems, Man and Cybernetics*, pages 551–556, 2003.

D. Y. Choi. Enhancing the power of web search engines by means of fuzzy query. *Decision Support Systems*, 35(1):31–44, 2003.

J. R. Conlon. Hope springs eternal: learning and the stability of cooperation in short horizon repeated games. *Journal of Economic Theory*, 112(1):35–36, 2003.

R. Cooley, B. Mobasher, and J. Srivastava. Data preparation for mining world wide web browsing patterns. *Knowledge and Information Systems*, 1(1):5–32, 1999.

D. R. Cox and H. D. Miller. *The Theory of Stochastic Processes*. Chapman and Hall, London, 1984.

J. Crawshaw and J. Chambers. *A Concise Course in Advanced Level Statistics*. Stanley Thornes Pub Ltd, 2001.

H. Cui, J. R. Wen, and W. Y. Ma. Query expansion by mining user logs. *IEEE Transactions on Knowledge and Data Engineering*, 15(4):829–839, 2003.

A. Damodaran. *Investment Valuation, 2nd edition*, chapter 6: Market Efficiency – Theory and Models. Wiley, 2001.

M. Daniels. Integrating simulation technologies with swarm. In *Proceedings of the Conference Agent Simulation: Applications, Models and Tools*, 1999.

DBE. Annex I: Description of Work, Digital Business Ecosystem. Technical report, 2002.

P. De Wilde. Fuzzy utility and equilibria. *IEEE Transactions on Systems, Man and Cybernetics, Part B*, 34(4):1774–1785, 2004.

P. De Wilde. Controlling the adaptation of a population of agents. In E. Alonso, editor, *Proceedings of Second Symposium on Adaptive Agents and Multi-agent Systems, London, April 4-5, 2002*, pages 75–80, Brighton, 2002.

P. De Wilde, H. S. Nwana, and L. C. Lee. Stability, fairness and scalability of multi-agent systems. *International Journal of Knowledge-Based Intelligent Engineering Systems*, 3(2):84–91, 1999a.

P. De Wilde, N. Sziribik, J. Goossenaerts, L. Correia, and R. Ribeiro. Annex 1: IST-1999-10304. Technical report, 1999b.

P. De Wilde, M. Chli, L. Correia, R. Ribeiro, P. Mariano, V. Abramov, and J. Goossenaerts. Adapting populations of agents. *Lecture Notes in Artificial Intelligence*, 2636:110–124, 2003.

M. D'Inverno, M. Luck, M. Georgeff, D. Kinny, and M. Wooldridge. The dMARS architecture: A specification of the distributed multi-agent reasoning system. *Autonomous Agents and Multi-agent Systems*, 9(1-2):5–53, 2004.

J. E. Doran, S. Franklin, N. R. Jennings, and T. J. Norman. On cooperation in multi-agent systems. *Knowledge Engineering Review*, 12(3):309–314, 1997.

M. Dorigo and L. M. Gambardella. Ant colony system: a cooperative learning approach to the traveling salesman problem. *IEEE Transactions on Evolutionary Computation*, 1(1):53–66, 1997.

E. N. Efthimiadis. Query expansion. *Annual Review of Information Systems and Technology*, 31:121–187, 1996.

T. Eguchi, K. Hirasawa, J. Hu, and N. Ota. Aa study of evolutionary multiagent models based on symbiosis. *IEEE Transactions on Systems, Man and Cybernetics, Part B*, 36(1):179–193, 2006.

T. R. Eikebrokk and D. H. Olsen. Co-opetition and e-business success in SMEs: An empirical investigation of european SMEs. In *Proceedings of the 38th Annual Hawaii International Conference on System Sciences (HICSS'05) – Track 7*, page 162.1, Washington, DC, USA, 2005. IEEE Computer Society.

E. J. Elton, M. J. Gruber, S. J. Brown, and W. N. Goetzmann. *Modern Portfolio Theory and Investment Analysis*. Wiley, 2006.

J. M. Epstein. Modeling civil violence: An agent-based computational approach. *Proceedings of the National Academy of Sciences, U.S.A.*, 99(10, Supplement 3): 7243–7250, 2002.

R. Falcone and C. Castelfranchi. The human in the loop of a delegated agent: the theory of adjustable social autonomy. *IEEE Transactions on Systems, Man and Cybernetics, Part A*, 31(5):406–418, 2001.

E. F. Fama. Efficient capital markets: A review of theory and empirical work. *Journal of Finance*, 25:383–417, 1970.

S. S. Fatima, M. Wooldridge, and N. R. Jennings. Sequential auctions for objects with common and private values. In *Proceedings of the Fourth International*

Joint Conference on Autonomous Agents and Multi-agent Systems, pages 635 – 642, 2005.

R. A Fisher. *Statistical methods for research workers, 14th ed.* Oliver and Boyd, 1970.

W. N. Francis and H. Kucera. Brown corpus manual. http://helmer.aksis.uib.no/icame/brown/bcm.html, 1979.

R. P. Gabriel and R. Goldman. Open source: beyond the fairytales, 2002. URL `http://opensource.mit.edu/papers/gabrielgoldman.pdf`. [Online; accessed 25-May-2005].

C. W. Gardiner. *Handbook of Stochastic Methods: For Physics, Chemistry and the Natural Sciences (Springer Series in Synergetics)*. Springer-Verlag, Berlin and Heidelberg, 2004.

S. Gauch and J. B. Smith. Search improvement via automatic query reformulation. *ACM Transactions on Information Systems*, 9(3):249–280, 1991. ISSN 1046-8188. doi: http://doi.acm.org/10.1145/125187.125193.

N. Glance, T. Hogg, and B. A. Huberman. Computational ecosystems in a changing environment. *International Journal of Modern Physics*, C(2):735–753, 1991.

D. E. Goldberg. *Genetic Algorithms in Search, Optimization and Machine Learning*. Addison-Wesley Longman Publishing Co., Inc., Boston, MA, USA, 1989. ISBN 0201157675.

S. Goonatilake and P. Treleaven. *Intelligent Systems for Finance and Business*. Wiley, New York, NY, USA, 1995. ISBN 0471944041.

S. Grand, G. von Krogh, D. Leonard, and W. Swap. Resource allocation beyond firm boundaries: A multi-level model for open source innovation. *Long Range Planning*, 37(6):591–610, 2004.

N. Griffiths and M. Luck. Coalition formation through motivation and trust. In *Proceedings of the Second International Joint Conference on Autonomous Agents and Multi-agent Systems*, pages 17–24, New York, NY, USA, 2003. ACM Press. ISBN 1-58113-683-8. doi: http://doi.acm.org/10.1145/860575.860579.

R. Groeneveld. Economic considerations in the optimal size and number of reserve sites. *Ecological Economics*, 52(2):219–228, 2005.

Z. Guo, H. K. Han, and J. C. Tay. Sufficiency verification of HIV-1 pathogenesis based on multi-agent simulation. In *GECCO '05: Proceedings of the 2005 Conference on Genetic and Evolutionary Computation*, pages 305–312, 2005.

J. Hagedoorn. Inter-firm R&D partnerships: an overview of major trends and patterns since 1960. *Research Policy*, 31(4):477–492, 2002.

B. Hamaide, C. S. ReVelle, and S. A. Malcolm. Biological reserves, rare species and the trade-off between species abundance and species diversity. *Ecological Economics*, in press, 2005.

L. Han, L. Sun, G. Chen, and L. Xie. ADSS: An approach to determining semantic similarity. *Advances in Engineering Software*, 37(2):129–132, 2006.

D. Harman. Towards interactive query expansion. In *Proceedings of the 11th annual international ACM SIGIR conference on Research and development in information retrieval*, pages 321–331. ACM Press, 1988. ISBN 2-7061-0309-4. doi: http://doi.acm.org/10.1145/62437.62469.

W. R. Hersh and P. Over. The TREC-2001 Interactive Track Report. In *Proceedings of the Tenth Text Retrieval Conference (TREC-10), Gaithersburg*, November 2001.

W. G. S. Hines. Evolutionary stable strategies: a review of basic theory. *Theoretical Population Biology*, 31:195–272, 1987.

J. H. Holland. Adaptation. *Progress in Theoretical Biology*, 4:263–293, 1976.

B. Horling and V. Lesser. Using ODML to model organizations for multi-agent systems. In *Proceedings of the International Conference on Intelligent Agent Technology (IAT 2005)*, pages 72–80, 2005.

J. Hou and Y. Zhang. Effectively finding relevant web pages from linkage information. *IEEE Transactions on Knowledge and Data Engineering*, 15(4):940–951, 2003.

C. Huang. Using intelligent agents to manage fuzzy business processes. *IEEE Transactions on Systems, Man and Cybernetics, Part A*, 31(6):508–523, 2001.

S. F. Huin. Managing deployment of ERP systems in SMEs using multi-agent systems. *International Journal of Project Management*, 22(6):511–517, 2004.

N. R. Jennings, K. Sycara, and M. Wooldridge. A roadmap of agent research and development. *Autonomous Agents and Multi-agent Systems*, 1(1):7–38, 1998.

Z. Jing, E. Billard, and S. Lakshmivarahan. Learning in multilevel games with incomplete information. ii. *IEEE Transactions on Systems, Man and Cybernetics, Part B*, 29(3):340–349, 1999.

J. P. Johnson. Open source software: private provision of a public good. *Journal of Economics and Management Strategy*, 11(4):637–662, 2002.

T. Kaihara. Supply chain management with market economics. *International Journal of Production Economics*, 73(1):5–14, 2001.

J. O. Kephart. Software agents and the route to information economy. *Proceedings of the National Academy of Sciences, U.S.A.*, 99(10, Supplement 3):7207–7213, 2002.

J. O. Kephart, J. E. Hanson, and J. Sairamesh. Price and niche wars in a free-market economy of software agents. *Artificial. Life*, 4(1):1–23, 1997. ISSN 1064-5462. doi: http://dx.doi.org/10.1162/106454698568413.

J. O. Kephart, J. E. Hanson, and A. R. Greenwald. Dynamic pricing by software agents. *Computer Networks*, 32:731–752, 2000.

H. K. Khalil. *Nonlinear Systems*. Prentice Hall; 2nd edition, 1995.

M. S. Khan and S. W. Khor. Web document clustering using a hybrid neural network. *Applied Soft Computing*, 4(4):423–432, 2004.

A. P. Kirman and N. J. Vriend. Evolving market structure: An ACE model of price dispersion and loyalty. *Journal of Economic Dynamics and Control*, 25(3):459–502, 2001.

R. Krovetz. Viewing morphology as an inference process. In *Proceedings of ACM SIGIR Conference*, pages 191–202, 1993.

B. LeBaron. Building the Santa Fe Artificial Stock Market. Working Paper, June 2002. Available at http://www.econ.iastate.edu/tesfatsi/blake.sfisum.pdf.

C. S. Lee and C. Y. Pan. An intelligent fuzzy agent for meeting scheduling decision support system. *Fuzzy Sets and Systems*, 142:467–488, 2004.

L. C. Lee, H. S. Nwana, D. T. Ndumu, and P. De Wilde. On the stability, scalability, and performance of multi-agent systems. *BT Technology Journal*, 16(3):94–103, 1998.

J. Lerner and J. Tirole. Some simple economics of open source. *Journal of Industrial Economics*, 50:197–234, June 2002.

V. R. Lesser. Multiagent systems: an emerging subdiscipline of AI. *ACM Computing Surveys (CSUR)*, 27(3):340–342, 1995.

A. Leuski. Evaluating document clustering for interactive information retrieval. In *Proceedings of the Tenth International Conference on Information and Knowledge Management*, pages 33–40. ACM Press, 2001. ISBN 1-58113-436-3. doi: http://doi.acm.org/10.1145/502585.502592.

C. Li, J.A. Giampapa, and K. Sycara. Bilateral negotiation decisions with uncertain dynamic outside options. *IEEE Transactions on Systems, Man, and Cybernetics, Part C: Special Issue on Game-theoretic Analysis and Stochastic Simulation of Negotiation Agents*, 36(1):31–44, 2006.

A. Likhodedov and T. Sandholm. Approximating revenue-maximizing combinatorial auctions. In *Proceedings of the National Conference on Artificial Intelligence*, 2005.

C. H. Loch and B. A. Huberman. A punctuated-equilibrium model of technology diffusion. *Management Science*, 45(2):160–177, 1999. ISSN 0025-1909.

C. de Loupy, P. Bellot, M. El-Beze, and P.F. Marteau. Query expansion and classification of retrieved documents. In *Proceedings of the Seventh Text REtrieval Conference (TREC-7)*, pages 382–389, 1998.

M. Luck and P. McBurney. Agent-based computing for next generation apps. *IT-NOW*, 47:24–25, 2005.

M. Luck, P. McBurney, and C. Preist. *Agent Technology: Enabling Next Generation Computing (A Roadmap for Agent Based Computing)*. AgentLink, 2003.

P. Maes. Agents that reduce work and information overload. *Commun. ACM*, 37(7): 30–40, 1994.

M. Magennis and C. J. van Rijsbergen. The potential and actual effectiveness of interactive query expansion. In *Proceedings of the 20th Annual International ACM SIGIR Conference on Research and Development in Information Retrieval*, pages 324–332. ACM Press, 1997. ISBN 0-89791-836-3. doi: http://doi.acm.org/10.1145/258525.258603.

P. Maldonado, C. Carrascosa, and V. Botti. Using negotiation techniques as time-restricted scheduling policies on intelligent agents. *Lecture Notes in Artificial Intelligence*, 3690:608–611, 2005.

P. Mariano, F. Buarque, M. Marques, M. Chli, and V. Abramov. Workpackage 7.1: Demonstrator (IST-1999-10304). Technical report, 2003.

M. Marques, P. Mariano, Luis Correia, and Ribeiro R. Workpackage 5: Adaptability (IST-1999-10304). Technical report, Universidade Nova de Lisboa, 2003.

T. P. Martin and B. Azvine. Acquisition of soft taxonomies for intelligent personal hierarchies and the soft semantic web. *BT Technology Journal*, 21(4):113–122, 2003.

M. J. Martin-Bautista, D. Sanchez, J. Chamorro-Martinez, J. M. Serrano, and M. A. Vila. Mining web documents to find additional query terms using fuzzy association rules. *Fuzzy Sets and Systems*, 148(1):85–104, 2004.

T. Marwala, P. De Wilde, L. Correia, P. Mariano, R. Ribeiro, V. Abramov, N. Szirbik, and J. Goossenaerts. Scalability and optimisation of a committee of agents using genetic algorithms. In *Proceedings of Fourth International ICSC Symposium Soft Computing and Intelligent Systems For Industry*, 2001.

A. Mas-Colell, M. D. Whinston, and J. R. Green. *Microeconomic Theory*. Oxford University Press Inc, USA, 2004.

P. McBurney and S. Parsons. A geometric semantics for dialogue game protocols for autonomous agent interactions. *Electronic Notes in Theoretical Computer Science*, 52(2):1–18, 2004.

G. A. Miller. WordNet: a lexical database for English. *Communications of the ACM*, 38(11):39–41, 1995. ISSN 0001-0782. doi: http://doi.acm.org/10.1145/219717.219748.

M. Mitra, A. Singhal, and C. Buckley. Improving automatic query expansion. In *Proceedings of the 21st Annual International ACM SIGIR Conference on Research and Development in Information Retrieval*, pages 206–214. ACM Press, 1998. ISBN 1-58113-015-5. doi: http://doi.acm.org/10.1145/290941.290995.

F. Mondada, D. Floreano, A. Guignard, J. L. Deneubourg, L. Gambardella, S. Nolfi, and M. Dorigo. Search for rescue: an application for the SWARM-BOT self-assembling robot concept. Technical Report LSA2-I2S-STI, Swiss Federal Institute of Technology, Lausanne, Switzerland, September 2002.

A. Moukas and P. Maes. Amalthaea: An evolving multi-agent information filtering and discovery system for the WWW. *Autonomous Agents and Multi-agent Systems*, 1(1):59–98, 1998.

Y.-E. Nahm and H. Ishikawa. A hybrid multi-agent system architecture for enterprise integration using computer networks. *Robotics and Computer-Integrated Manufacturing*, 21(3):217–234, 2005.

R. Narula. R&D collaboration by SMEs: new opportunities and limitations in the face of globalisation. *Technovation*, 24(2):153–161, 2004.

T. D. Nguyen and N. R. Jennings. Managing commitments in multiple concurrent negotiations. *International Journal of Electronic Commerce Research and Applications*, 4(4):362–376, 2005.

M. Nikravesh, V. Loia, and B. Azvine. Fuzzy logic and the Internet (FLINT): Internet, World Wide Web and search engines. *Soft Computing*, 6:287–299, 2002.

G. Nitschke. Emergence of cooperation: State of the art. *Artificial Life*, 11(3): 367–396, 2005.

J. R. Norris. *Markov Chains*. Cambridge University Press, 1997.

S. Olafsson. On the stability of strategies in competitive systems. *International Journal of Systems Science*, 26:1289–1312, 1995.

L. Perko. *Differential Equations and Dynamical Systems*. Springer-Verlag, New York, 2001.

J. Piaget. *Six études de psychologie*. Folio essais. Denoël, Paris, 1964.

M. F. Porter. An algorithm for suffix stripping. *Program: Automated Library and Information Systems*, 14(3):130–137, 1980.

J. L. Ribeiro Filho, P. C. Treleaven, and C. Alippi. Genetic-algorithm programming environments. *Computer*, 27(6):28–43, 1994.

S. E. Robertson, C. L. Thompson, M. J. Macaskill, and J. D. Bovey. Weighting, ranking and relevance feedback in a front-end system. *Journal of Information Science*, 12(1-2):71–75, 1986. ISSN 0165-5515.

M. R. Rodrigues, A. C. da Rocha Costa, and R. H. Bordini. A system of exchange values to support social interactions in artificial societies. In *AAMAS '03: Proceedings of the Second International Joint Conference on Autonomous Agents and Multi-agent Systems*, pages 81–88, 2003.

J. S. Rosenschein and G. Zlotkin. *Rules of Encounter: Designing Conventions for Automated Negotiation among Computers.* MIT Press, Cambridge, Massachusetts, 1994.

S. M. Ross. *Stochastic Processes (Wiley Series in Probability and Statistics).* Wiley, Hoboken, New Jersey, 1996.

I. Ruthven. Re-examining the potential effectiveness of interactive query expansion. In *Proceedings of the 26th Annual International ACM SIGIR Conference on Research and Development in Information Retrieval*, pages 213–220. ACM Press, 2003. ISBN 1-58113-646-3. doi: http://doi.acm.org/10.1145/860435.860475.

J. Sairamesh and J. O. Kephart. Price dynamics and quality in information markets. *Decision Support Systems*, 28:35–47, 2000.

G. Salton and C. Buckley. Term-weighting approaches in automatic text retrieval. *Information Processing and Management*, 24(5):513–523, 1988.

M. Sanderson. Retrieving with good sense. *Information Retrieval*, 2:47–67, 1999.

T. Sandholm and V. R. Lesser. Coalitions among computationally bounded agents. *Artificial Intelligence*, 94(1-2):99–137, 1997.

R. Savit, R. Manuca, and R. Riolo. Adaptive competition, market efficiency and phase transitions. *Physical Review Letters*, 82(10):2203–2206, 1999.

M. Schillo, H. J. Burckert, K. Fischer, and M. Klusch. Towards a definition of robustness for market-style open multi-agent systems. In *Proceedings of the Fifth International Conference on Autonomous Agents*, pages 75–76, 2001.

K. Schmidt and M. Schnitzer. Public Subsidies for open source; Some economic policy issues of the software market. Technical Report 3793, C.E.P.R. Discussion Papers, February 2003. Available at http://ideas.repec.org/p/cpr/ceprdp/3793.html.

J. Shen, I. Weber, and V. Lesser. OAR: A Formal Framework for Multi-agent Negotiation. *Proceedings of the Twentieth National Conference on Artificial Intelligence (AAAI 05)*, 2005.

K. M. Sim and E. Wong. Toward market-driven agents for electronic auction. *IEEE Transactions on Systems, Man and Cybernetics, Part A*, 31(6):474–484, 2001.

M. Simoes-Marques, P. Mariano, R. Ribeiro, L. Correia, M. Chli, P. De Wilde, V. Abramov, and J. Goossenaerts. Contributions to adaptable agent societies. In *Proceedings of the Ninth IEEE International Conference on Emerging Technologies and Factory Automation*, 2003.

K. Sparck-Jones. *Automatic Keyword Classification for Information Retrieval.* Butterworth, London, UK, 1971. ISBN 0408701374.

W. M. Spears. *Evolutionary Algoriths.* Springer, 2000.

A. Spink, D. Wolfram, B.J. Jansen, and T. Saracevic. Searching the web: the public and their queries. *Journal of the American Society for Information Science and Technology*, 52(3):226–234, 2001.

D. Stirzaker. *Stochastic Processes and Models.* Oxford University Press, UK, 2005.

S. Sumathi, T. Hamsapriya, and P. Surekha. *Evolutionary Intelligence: An Introduction to Theory and Applications with Matlab.* Springer, 2008.

M. Sysi-Aho, A. Chakraborti, and K. Kaski. Searching for good strategies in adaptive minority games. *Physical Review E*, 69(3):36125–1–36125–7, 2004.

J. M. Tenenbaum and R. Khare. Business services networks: delivering the promises of B2B. In *BSN '05: Proceedings of the IEEE EEE05 International Workshop on Business Services Networks*, pages 8–8, Piscataway, NJ, USA, 2005. IEEE Press.

G. J. Tesauro and J. O. Kephart. Foresight-based pricing algorithms in agent economies. *Decision Support Systems*, 28:49–60, 2000.

L. Tesfatsion. *Handbook of Computational Economics, Vol. 2: Agent-Based Computational Economics*, chapter 1, Agent-Based Computational Economics: A Constructive Approach to Economic Theory. North-Holland, 2005.

J. D. Thomas and K. Sycara. Heterogeneity, stability, and efficiency in distributed systems. In *Proceedings of the Third International Conference on Multi-agent Systems*, pages 293–300, 1998.

H. Van Dyke Parunak, J. Sauter, and S. Clark. Toward the specification and design of industrial synthetic ecosystems. In *Proceedings of the Fourth International Workshop on Agent Theories, Architectures, and Languages*, 1997.

H. Van Dyke Parunak, S. Brueckner, and R. Savit. Universality in multi-agent systems. In *AAMAS '04: Proceedings of the Third International Joint Conference on Autonomous Agents and Multi-agent Systems*, pages 930–937, Washington, DC, USA, 2004. IEEE Computer Society. ISBN 1-58113-864-4. doi: http://dx.doi.org/10.1109/AAMAS.2004.290.

O. Vechtomova, S. Robertson, and S. Jones. Query expansion with long-span collocates. *Information Retrieval*, 6(2):251–273, 2003.

J. M. Vidal. The effects of cooperation on multiagent search in task-oriented domains. In *Proceedings of the First International Joint Conference on Autonomous Agents and Multi-agent Systems*, pages 453–454, 2002.

R. Vidal, O. Shakernia, J. Kim, D. Shim, and S. Sastry. Probabilistic pursuit-evasion games: Theory, implementation and experimental evaluation. *IEEE Transactions on Robotics and Automation*, 18(5):662–669, 2002.

E. M. Voorhees. Using WordNet to disambiguate word senses for text retrieval. In *Proceedings of the 16th Annual International ACM SIGIR Conference on Research and Development in Information Retrieval*, pages 171–180. ACM Press, 1993. ISBN 0-89791-605-0. doi: http://doi.acm.org/10.1145/160688.160715.

F. Wang. A fuzzy grammar and possibility theory-based natural language user interface for spatial queries. *Fuzzy Sets and Systems,*, 113(1):147–159, 2000.

J. W. Weibull. *Evolutionary Game Theory.* MIT Press, Cambridge, MA, USA, 1995.

R. White, J. Jose, and I. Ruthven. Comparing explicit and implicit feedback techniques for web retrieval: TREC-10 interactive track report. In *Proceedings of the Tenth Text Retrieval Conference (TREC-10), Gaithersburg*, November 2001.

S. Winkler, M. Affenzeller, and S. Wagner. Advanced genetic programming based machine learning. *Journal of Mathematical Modelling and Algorithms*, 6(3):455–480, 2007.

T. Yamasaki and T. Ushio. An application of a computational ecology model to a routing method in computer networks. *IEEE Transactions on Systems, Man and Cybernetics, Part B*, 32(1):99–106, 2002.

Y. Zhang, H. Ghenniwa, and W. M. Shen. Agent-based personal assistance in collaborative design environments. *Lecture Notes in Computer Science*, 3865:284–293, 2006.

G. Zlotkin and J. S. Rosenschein. Mechanism design for automated negotiation, and its application to task oriented domains. *Artificial Intelligence*, 86:195–244, 1996.

Index

Printed in the United States
148239LV00002B/18/P

9 781848 820623